£40

WITHDRAWN
from
STIRLING UNIVERSITY LIBRARY

Cambridge astrophysics series

Pulsar astronomy

In this series

1. Active Galactic Nuclei
   edited by C. Hazard and S. Mitton
2. Globular Clusters
   edited by D. A. Hanes and B. F. Madore
3. Low Light-level Detectors in Astronomy
   by M. J. Eccles, M. E. Sim and K. P. Tritton
4. Accretion-driven Stellar X-ray Sources
   edited by W. H. G. Lewin and E. P. J. van den Heuvel
5. The Solar Granulation
   by R. J. Bray, R. E. Loughhead and C. J. Durrant
6. Interacting Binary Stars
   edited by J. E. Pringle and R. A. Wade
7. Spectroscopy of Astrophysical Plasmas
   by A. Dalgarno and D. Layzer
8. Accretion Power in Astrophysics
   by J. Frank, A. R. King and D. J. Raine
9. Gamma-ray Astronomy
   by P. V. Ramana Murthy and A. W. Wolfendale
10. Quasar Astronomy
    by D. W. Weedman
11. X-ray Emission from Clusters of Galaxies
    by C. L. Sarazin
12. The Symbiotic Stars
    by S. J. Kenyon
13. High Speed Astronomical Photometry
    by B. Warner
14. The Physics of Solar Flares
    by E. Tandberg-Hanssen and A. G. Emslie
15. X-ray Detectors in Astronomy
    G. W. Fraser
16. Pulsar Astronomy
    A. G. Lyne and F. Graham-Smith
17. Molecular Collisions in the Interstellar Medium
    D. Flower

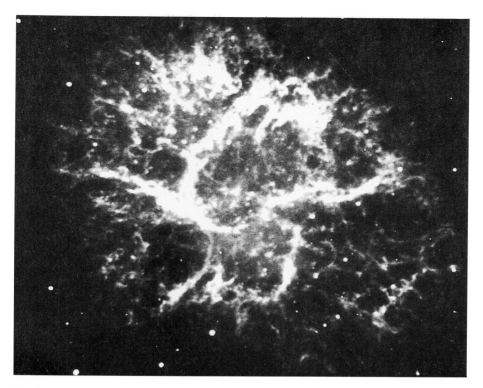

The Crab Nebula in hydrogen-alpha light. This CCD picture was obtained with the Kapteyn Telescope on La Palma by D. H. P. Jones of the Royal Greenwich Observatory. (North is at the top.)

PULSAR ASTRONOMY

A. G. LYNE
Reader in Radio Astronomy,
Department of Physics, University of Manchester

F. GRAHAM-SMITH
Astronomer Royal and Langworthy Professor of Physics,
University of Manchester

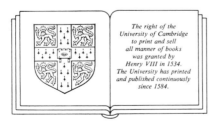

CAMBRIDGE UNIVERSITY PRESS
Cambridge
New York   Port Chester
Melbourne   Sydney

Published by the Press Syndicate of the University of Cambridge
The Pitt Building, Trumpington Street, Cambridge CB2 1RP
40 West 20th Street, New York, NY 10011, USA
10 Stamford Road, Oakleigh, Melbourne 3166, Australia

© Cambridge University Press 1990

First published 1990

Printed in Great Britain at the University Press, Cambridge

British Library cataloguing in publication data
Lyne, A. G.
Pulsar astronomy.
1. Pulsars
I. Title II. Graham-Smith, Francis, *1923–*
523

Library of Congress cataloguing in publication data
Pulsar astronomy / A. G. Lyne, F. Graham-Smith
p. cm. – (Cambridge astrophysics series)
Bibliography: p.
Includes index.
ISBN 0 521 32681 8
1. Pulsars. 2. Stars, Binary. I. Graham-Smith, Francis, Sir.
1923– II. Title. III. Series.
QB843.P8L86 1990
523.8'874 – dc20   89-35679   CIP

ISBN 0 521 32681 8

Contents

Preface	*page*	xiii
1 The discovery of pulsars		1
1.1 Neutron stars		1
1.2 The radio discovery		2
1.3 Interplanetary scintillation		4
1.4 The *Nature* letter of February 1968		6
1.5 The identification with neutron stars		7
1.6 Oscillations and orbits		8
1.7 Oscillations		9
1.8 Planetary and binary orbits		9
1.9 Rotation and slowdown		10
1.10 Optical pulses from the Crab Pulsar		12
1.11 X-ray pulses from the Crab Pulsar		13
1.12 The development of pulsar research		15
2 The pulsars – an overview		17
2.1 Periods		17
2.2 Galactic population		18
2.3 The structure of neutron stars		18
2.4 The magnetic fields of neutron stars		20
2.5 Radio pulse characteristics		23
2.6 Optical and other high energy emission		24
2.7 The interstellar medium		25
3 Searches and surveys		26
3.1 The problem of sensitivity		26
3.2 Frequency dispersion in pulse arrival time		27
3.3 Search techniques		28
3.4 Search for periodic pulses		31
3.5 The pulsar surveys		33
3.6 Selection effects – the limits of the surveys		35
3.7 Searches for millisecond pulsars		36

4 The distances of the pulsars — 38
4.1 Introduction — 38
4.2 Pulsar distances from parallax — 38
4.3 Pulsar distances from neutral hydrogen absorption — 39
4.4 Optically identified pulsars — 41
4.5 Interstellar hydrogen — 42
4.6 The H II regions — 43
4.7 Model electron distribution — 45

5 Pulse timing — 46
5.1 Pulsar positions and the Earth's orbit — 46
5.2 The barycentric correction — 49
5.3 The General Relativistic correction — 50
5.4 The fundamental reference frame — 51
5.5 Period changes — 51
5.6 Pulsar ages and the braking index — 52
5.7 Pulsars as standard clocks — 54
5.8 Proper motion — 56
5.9 Binary pulsar systems — 57
5.10 The binary pulsar PSR 1913+16 — 58
5.11 Relativistic effects in the binary orbit — 60

6 Timing irregularities — 63
6.1 Timing noise and glitches — 63
6.2 The Vela Pulsar PSR 0833−45 — 64
6.3 The Crab Pulsar PSR 0531+21 — 67
6.4 Timing discontinuities in other pulsars — 69
6.5 Timing noise — 72
6.6 Adjustments in ellipticity: starquakes — 74
6.7 Two-component models: the glitch function — 75
6.8 Vorticity in the neutron fluid — 76
6.9 Vortex lattice oscillation — 78

7 The young pulsars — 80
7.1 The Crab Pulsar PSR 0531+21 — 81
   7.1.1 Discovery — 81
   7.1.2 Pulse profile — 82
   7.1.3 Spectrum — 85
   7.1.4 Polarisation — 87
   7.1.5 Intensity fluctuations – the giant pulses — 87
   7.1.6 Intensity fluctuations – long-term scintillations — 89
   7.1.7 Variable radio pulse broadening — 89
   7.1.8 Crab pulse timing — 91
7.2 The Vela Pulsar PSR 0833−45 — 93
   7.2.1 Discovery — 93
   7.2.2 Pulse profile — 93
   7.2.3 The spectrum — 95
   7.2.4 Polarisation of the integrated radio profile — 95
   7.2.5 Vela pulse timing — 95

7.3	PSR 1509−58	96
7.4	Other young pulsars	98

8 The Galactic population of pulsars — 100

8.1	The surveys	100
8.2	The observed distribution	101
8.3	The derived luminosity and spatial distributions	101
8.4	Pulsar velocities and ages	105
8.5	The pulsar birthrate	108
8.6	The population of millisecond pulsars	110
8.7	The galactic centre region	110
8.8	The origin in supernovae	110

9 Supernovae — 112

9.1	The nature of supernovae	112
9.2	Stellar collapse	114
9.3	Luminosity decay	116
9.4	Supernovae in binary systems	116
9.5	Quiet collapse	118
9.6	Frequency of occurrence of supernovae	120
9.7	Associations between pulsars and supernovae	121
9.8	The Crab Nebula	122
9.9	The continuum radiation from the Crab Nebula	124
9.10	The energy supply	125

10 Binary and millisecond pulsars — 128

10.1	A separate population	128
10.2	Circular and elliptical orbits	128
10.3	The globular cluster pulsars	132
10.4	The eclipsing millisecond binary	133
10.5	The masses of the binary pulsars and their companions	134
10.6	Evolution of binary systems	136
10.7	Stellar evolution	137
10.8	Mass transfer	137

11 The X-ray binaries and bursters — 141

11.1	Optical companions of X-ray binaries	142
11.2	The massive X-ray binaries (MXBs)	143
11.3	The low-mass X-ray binaries (LMXBs)	144
11.4	Spin-up, accretion and inertia	146
11.5	X-ray light curves and a spectral line	148
11.6	The X-ray bursters	149
11.7	The Rapid Burster	151
11.8	Magnetic field decay	152
11.9	Quasi-periodic oscillations	152

12 Integrated pulse profiles — 154

12.1	The integrated profiles	155
12.2	Radio frequency dependence	159
12.3	Extended profiles and interpulses	160

12.4	The overall angular width	162
12.5	Circular polarisation	164
12.6	Orthogonal polarisation	165
12.7	An empirical model of polar cap emission	165

13 Individual pulses — 168

13.1	Single pulse intensities and pulse nulling	170
13.2	Mode changing	172
13.3	Sub-pulses	173
13.4	Drifting and modulation	174
13.5	Drift rates	177
13.6	Drifting after nulling	179
13.7	The polarisation of sub-pulses	181
13.8	Microstructure	181

14 Geometry of the emitting regions — 184

14.1	The outer magnetosphere gap	184
14.2	Non-orthogonal configurations	186
14.3	Polarisation of the high-energy source	187
14.4	The polar cap and the source of radio emission	189

15 Radiation processes — 192

15.1	Cyclotron radiation	192
15.2	Synchrotron radiation	194
15.3	Curvature radiation	197
15.4	The synchrotron spectrum	198
15.5	Self absorption	199
15.6	Inverse Compton radiation	199
15.7	Maser amplification	200
15.8	Coherence in the radio emission	200
15.9	Relativistic beaming	201

16 The pulsar emission mechanisms — 203

16.1	The outer magnetosphere gaps	203
16.2	The Crab Pulsar: high energy emission	205
16.3	The infrared spectrum	207
16.4	The Crab Pulsar double radio pulse	208
16.5	The Vela Pulsar high energy pulses	208
16.6	The radio pulses	209

17 Interstellar scintillation and scattering — 211

17.1	A thin screen model	212
17.2	Diffraction theory of scintillation	213
17.3	Thick (extended) scattering screen	215
17.4	The Fresnel distance	215
17.5	Distribution in size of the irregularities	216
17.6	Dynamic spectra	217
17.7	Frequency drifting	217
17.8	The velocity of the scintillation pattern	221

17.9	Pulse broadening	223
17.10	Multiple scattering	224
17.11	Observations of pulse broadening	225
17.12	Apparent source diameters	228
17.13	Long-term intensity variations	229
18	**The interstellar magnetic field**	232
18.1	Optical and radio observations	232
18.2	Faraday rotation in pulsars	234
18.3	The configuration of the local field	235
19	**Achievements and prospects**	237
19.1	The population and the birthrate	237
19.2	Pulsar searches	239
19.3	Condensed matter within pulsars	239
19.4	Young pulsars and supernova remnants	240
19.5	The interstellar medium	242
19.6	The emission mechanism	243
20	**The Pulsar Catalogue**	244
	References	260
	Index	272

Preface

The two decades that have elapsed since their discovery have seen pulsars take their place in a surprisingly wide range of astrophysics. Not only are they the seat of some of the most extreme physical phenomena; they are now seen to be the end product of the most dramatic astrophysical events, the supernova explosions and the mass transfer, which plays such an important part in binary star systems. Furthermore, they are increasingly seen as a major tool in the exploration of interstellar space. Their discovery, by Anthony Hewish and Jocelyn Bell in 1967, is now seen to rank with the discovery of quasars and the microwave background as one of the major advances in modern astronomy.

When the news of the discovery reached Jodrell Bank, the authors were well placed to contribute, particularly by using the 250-ft Lovell radio telescope to investigate properties of the pulsar emission, such as its polarisation, intensity variations, and pulse drifting, and to make some of the earliest observations of interstellar scintillation. Searches for pulsars have now led to close collaboration with Australian colleagues, particularly in using the 210-ft Parkes radio telescope for searches in the southern hemisphere. The outstanding advance has been the establishment of an international network of observers and theorists who are closely in touch through a series of conferences and, even more potently, through electronic mail. This book is not needed by the members of this privileged circle; to them we offer our appreciation of their cooperation and our apologies for any misrepresentations of their work.

Our intention is that this book should be useful to a wider range of scientists, perhaps to students entering the field and perhaps to those who enjoy an exciting story of discovery and of the application of basic physics to a realm unattainable in earthbound laboratories.

Part of this book is based on an earlier monograph (*Pulsars*, by F. G.

Smith, Cambridge University Press 1977). Further and more detailed material may be found in IAU Symposia Nos. 95 on *Pulsars* (1980) (Reidel) and 125 on *The Origin and Evolution of Neutron Stars* (1987) (Kluwer Academic Publishers).

1

The discovery of pulsars

1.1 Neutron stars

In 1934, two astronomers, Walter Baade and Fritz Zwicky, proposed the existence of a new form of star, the neutron star, which would be the end point of stellar evolution. They wrote:

> with all reserve we advance the view that a super-nova represents the transition of an ordinary star into a neutron star, consisting mainly of neutrons. Such a star may possess a very small radius and an extremely high density.

These prophetic remarks seemed at the time to be beyond any possibility of actual observation, since a neutron star would be small, cold and inert, and would emit very little light. More than thirty years later the discovery of the pulsars, and the realisation a few months later that they were neutron stars, provided a totally unexpected verification of the proposal.

The physical conditions inside a neutron star are very different from laboratory experience. Densities up to 10^{14} g cm^{-3}, and magnetic fields up to 10^{12} gauss (10^8 tesla), are found in a star of solar mass but only about 30 kilometres in diameter. Again, predictions of these astonishing conditions were made before the discovery of pulsars. Oppenheimer and Volkoff in 1939 used a simple equation of state to predict the total mass, the density and the diameter; Hoyle, Narlikar and Wheeler in 1964 argued that a magnetic field of 10^{10} gauss might exist on a neutron star at the centre of the Crab Nebula; Pacini in 1967, just before the pulsar discovery, proposed that the rapid rotation of a highly magnetised neutron star might be the source of energy in the Crab Nebula.

Radio astronomers did not, however, set out to investigate the possibility that such bizarre objects might have detectable radio emission. No prediction had been made of the extremely powerful lighthouse beam of radio waves, producing radio pulses as the rotation of the neutron star sweeps the beam across the observer's line of sight. The observation of an astonishing and remarkably regular series of pulses was made by radio astronomers who were unfamiliar with the new theoretical concepts and who naturally took some time to connect their observations with predictions concerning some apparently unobservable objects.

Condensed stars, either white dwarfs or neutron stars, were predicted to be observable sources of X-rays. In 1964 independent predictions were made by Zel'dovich and Guseynov and Hayakawa and Matsouka, introducing the concept of binary star systems as X-ray sources. If in a binary star system one star is a condensed object and the other is a more massive normal star that is losing mass through a stellar wind, there might be a very large rate of accretion onto the condensed star, and a hot, dense, atmosphere would then develop. This atmosphere would radiate thermal X-rays.

The first X-ray astronomical observations were of the Sun and of the Crab Nebula. A more powerful source, Sco X–1, was a surprise discovery (Giacconi *et al.* 1962). This was soon identified with a visible star whose spectrum fitted that of a binary partner undergoing mass loss. A much greater surprise came from the X-ray satellite UHURU, launched in 1970. Two new pulsating sources were discovered, catalogued as Hercules X–1 and Centaurus X–3 (Giacconi *et al.* 1971). The explanation combined the neutron star concepts, already confirmed by the radio discoveries, with the binary star proposals, opening a new field in astrophysics.

1.2 The radio discovery

The spectacular growth of radio astronomy during the 30 years following the Second World War was marked by the introduction of a series of new observational techniques, each of which opened new fields of research. Each advance in technique was applied initially to a specific problem, but such was the richness of the radio sky that each new technique guided the observers into unexpected directions. Examples of such serendipity are provided by most of the major discoveries in radio astronomy. An investigation of the radio background, undertaken by J. S. Hey and his colleagues as an extension of their meteor radar work, yielded as a complete surprise the first discrete radio source, Cygnus A. New techniques in radio telescopes, designed in Cambridge to extend the counts of extragalactic sources for cosmological investigations, led unexpectedly to the discovery of quasars; these emerged from measurements at Jodrell Bank of the diameters of the unidentified extragalactic sources. The identification of quasars, itself a surprise, came from the development of lunar occultation and its use with a new radio telescope in

1.2 The radio discovery

Sydney. The added bonus, which so often follows an adventure into a new observational technique, has its example *par excellence* in the discovery of the pulsars.

At the start of the story we may ask why it was that pulsars were not discovered earlier than 1967. Their signals are very distinctive and often quite strong, so that, for example, the 250-ft radio telescope at Jodrell can be used to produce audible trains of pulses from several pulsars. The possibility of discovery had existed for ten years before it became reality. In fact, it turned out that pulsar signals had been recorded but not recognised when this telescope was used for a survey of background radiation several years before the actual discovery. The pulsar, now known as PSR 0329+54, left a clear imprint on several of the survey recordings. A similar story can be told for radio pulses from the planet Jupiter. These were discovered in 1954, although recordings made in Australia five years previously contained signals from Jupiter, which proved to be useful in subsequent analyses of the rotation period of the radio sources on the planet. An even more remarkable pre-discovery recording exists for the X-ray pulses from the Crab Pulsar. These were found on records from a balloon flight, which pre-dated the actual discovery rocket flights by two years. The signals were recorded, but not recognised.

The initial difficulty in recognition of the pulsar radio signals was that radio astronomers were not expecting to find rapid fluctuations in the signals from any celestial source. An impulsive radio signal received by a radio telescope was regarded as interference, generated in the multitudinous terrestrial impulsive sources, such as electrical machinery, power line discharges and automobile ignition. Indeed, most radio receivers were designed to reject or smooth out impulsive signals and to measure only steady signals, averaged over several seconds of integration time. Even if a shorter integrating time constant was in use, a series of impulses appearing on a chart recorder would excite no comment; interference of such regular appearance is to be expected, and is often encountered from such a simple device as an electric cattle fence on a farm within a mile or two of the radio telescope.

Two attributes were lacking in the apparatus used in these previous surveys: a short response time and a repetitive observing routine, which would show that the apparently sporadic signals were in fact from a permanent celestial source. These were both features of the survey of the sky for radio scintillation designed by A. Hewish, in the course of which the first pulsar was discovered.

1.3 Interplanetary scintillation

The familiar twinkling of visible stars, due to random refraction in the terrestrial atmosphere, has three distinct manifestations in radio astronomy, in which the refraction is due to ionised gas. Random refraction, causing scintillation of radio waves from celestial sources, occurs in the terrestrial ionosphere, in the ionised interplanetary gas in the solar system, and in the ionised interstellar gas of the Galaxy. In all three regions the radio waves from a distant point source traverse a medium with fluctuations of refractive index sufficient to deviate radio rays into paths which cross before they reach the observer, giving rise to interference and hence to variations in signal strength. All three types of radio twinkling, or 'scintillation', were discovered and investigated at Cambridge, and in all three investigations Hewish played a key part. Coincidentally, the theory of interstellar scintillation is important for pulsars, which now provide its most dramatic demonstration; the coincidence is that it was an investigation of interplanetary scintillation that led to the discovery of pulsars, even though the discovery was a by-product rather than the purpose of the investigation.

Hewish was working with a research student, Miss Jocelyn Bell (now Dr Bell-Burnell). They constructed a large receiving antenna for a comparatively long radio wavelength, 3.7 m, making a radio telescope which was sensitive to weak discrete radio sources. At this long wavelength the interplanetary scintillation effects are large, but they only occur for radio sources with a very small angular diameter. Scintillation is therefore seen as a distinguishing mark of the quasars, since the larger radio galaxies do not scintillate; Hewish later used the results of a survey with this system to study the distribution and population of these very distant extragalactic sources. The observational technique involved a repeated survey of the sky, using a receiver with a short time constant, which would follow the scintillation fluctuations.

The discovery was made by Jocelyn Bell within a month of the start of regular recordings in July 1967. Large fluctuations of signal were seen at about the same time on successive days. The characteristics of the signal looked unlike scintillation, and very like terrestrial interference. Hewish at first dismissed them as interference, such as might be picked up from a passing motor car. For several nights no signals appeared; as we now know, this must have been due to the random interstellar scintillation effects. Then they re-appeared, and continued to re-appear spasmodically. It soon became clear that the fluctuations were occurring four minutes earlier each day, as expected for a signal of celestial origin observed with a transit telescope and, in October, Hewish concluded that

1.3 Interplanetary scintillation

something new had turned up. What sort of celestial source could this be? He and his colleagues then used a recorder with an even faster response time and, in November, they first saw the amazingly regular pulses having a repetition period of about 1.337 seconds. Could they be man-made? Possibly they originated on a space-craft? Possibly they were the first radio signals from an extraterrestrial civilisation? (Fig. 1.1)

Fig. 1.1. Discovery observations of the first pulsar. (*a*) The first recording of PSR 1919+21; the signal resembled the radio interference also seen on this chart: (*b*) Fast chart recording showing individual pulses as downward deflections of the trace.

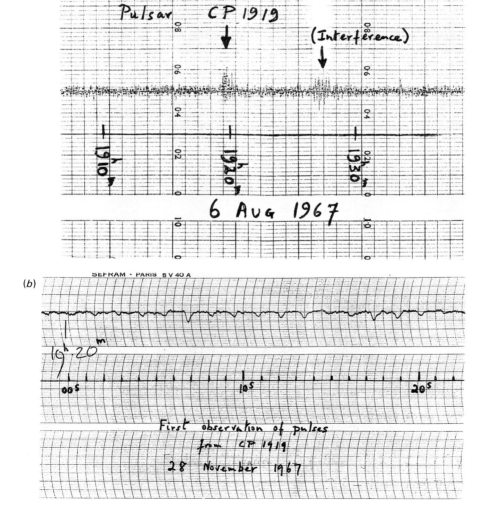

The last possibility was disturbing. If it became known to the public that signals were being received that might have come from intelligent extra-terrestrial sources – the 'little green men' of science fiction – the newspaper reporters would descend in strength on the observatory and destroy any chance of a peaceful solution to the problem. So there was intense activity but no communication for two months until, in February 1968, a classic paper appeared in *Nature* (Hewish *et al.* 1968).

1.4 The *Nature* letter of February 1968

The announcement of the discovery contained a remarkable analysis of the pulsating signal, which already showed that the source must lie outside the Solar System, and probably at typical stellar distances; furthermore, the rapidity of the pulsation showed that the source must be very small, and probably some form of condensed star, presumably either a white dwarf or a neutron star. These results have, of course, been expanded and overtaken by later work, but it should not be forgotten that, in the completely open and speculative atmosphere of the first few weeks after the discovery, the right conclusions were reached on the most important properties of the pulsating source. The location outside the Solar System came from observations of the Doppler effect of the Earth's motion on the pulse periodicity; this phenomenon also led to a positional determination. Admittedly, the first observations miscounted by one the number of pulses in one day, giving an error in period of one part in 6×10^4; but this seems to be the only subsequent correction to the pulse timing. It is particularly interesting to see that the paper specifically mentions a neutron star as a possible origin, when at that time the existence of neutron stars was only hypothetical. Indeed, the flow of speculative theoretical papers that was let loose by the discovery did not even follow up this idea at first, exploring instead every possible configuration of the more familiar binary systems and white dwarf stars.

A few days before the *Nature* letter appeared, the discovery was discussed at a colloquium in Cambridge. The news spread rapidly and radio astronomers immediately turned their attention to confirming the remarkable results. Only a fortnight separated the first paper and a *Nature* letter from Jodrell Bank giving some remarkable extra details of the radio pulses from this first pulsar, now known as PSR 1919+21. (PSR stands for Pulsating Source of Radio; the numbers refer to its position.) New discoveries of pulsars were made and announced by other observatories within a few months. By the middle of the year, significant contributions were being made by at least eight radio observatories.

1.5 The identification with neutron stars

This celebration of the discovery of pulsars was somewhat marred by allegations that Cambridge had withheld publication of its results instead of making all information freely available at the moment of discovery. The trouble seems to have been caused by a statement in the original *Nature* announcement that three other pulsars had been detected, and that their characteristics were being investigated. The statement was evidently made merely as supporting evidence for the astrophysical interpretations advanced in the *Nature* letter but it led to a bombardment of requests for advance information on the location and periodicities of these three further pulsars. Hewish refused to give further details until his measurements were complete; the results were published in *Nature* in April. His action was entirely in accordance with normal scientific protocol but it was misinterpreted as a deliberate obstruction by some would-be observers.

1.5 The identification with neutron stars

The historian of science will also enjoy the story of the theoretical papers which led to the identification of pulsars with neutron stars. It should be remembered that white dwarf stars were already observable and understood, while the further stage of condensation represented by a neutron star existed in a theory familiar only to certain astrophysicists who were concerned with highly condensed states of matter. Suggestions based on the more familiar white dwarf stars, and particularly on their various possible modes of oscillation, poured out from the theorists. Unknown to them, and apparently also unnoticed by Hewish, Dr Franco Pacini had already published a paper containing the solution to the nature of pulsars, again in *Nature*, and only a few months before the discovery. This was the paper (Pacini 1967) in which he showed that a rapidly rotating neutron star, with a strong dipolar magnetic field, would act as a very energetic electric generator, which could provide a source of energy for radiation from a surrounding nebula, such as the Crab Nebula. His work, and the original proposal by Baade and Zwicky, pointed the way to the subsequent discovery of the Crab Pulsar in the centre of the Nebula.

In June 1968, *Nature* published a letter from Professor T. Gold, of Cornell University, which set out very clearly the case for identifying the pulsars with rotating neutron stars. Between them, the two papers from Pacini and Gold contained the basic theory and the vital connection with the observations. The remarkable part of the story is that the two men were working in offices practically next door to one another at the time of Gold's paper, since Pacini was visiting Cornell University; nevertheless Gold did not even know of Pacini's earlier work, and there is no reference to it in his paper (Gold 1968). Collaboration was, of course, soon established, as may be seen from a paper from Pacini only a month later (Pacini 1968). These two men should clearly share the credit for the linkage between pulsars and neutron stars.

The confusion of theories persisted until the end of 1968, even though the correct theory had been clearly presented. Unfamiliarity with the concept of a neutron star seems to have been the main barrier to understanding, at least for the observers; it is interesting to see that both Hewish and Smith wrote forewords to a collection of *Nature* papers towards the end of 1968 in which they favoured explanations involving the more conventional white dwarf stars. The issue was settled dramatically by the discoveries of the short-period pulsars now known as the Vela and the Crab Pulsars. The experimental test was simple: theories involving white dwarf stars might account for pulsars with periods of about 1 s, and possibly even for ¼ s, the shortest period then known, but the Vela Pulsar discovered in Australia by Large, Vaughan and Mills (1968), had a period of only 89 ms, while the Crab Pulsar, discovered in the USA by Staelin and Reifenstein (1968) had the even shorter period of 33 ms. Only a neutron star could vibrate or rotate as fast as 30 times per second. Furthermore, as pointed out by Pacini and Gold, a rotation would slow down, but a vibration would not. Very soon a slowdown was discovered in the period of the Crab Pulsar (Richards and Comella 1969), and the identification with a neutron star was then certain. Furthermore, both the Crab and the Vela Pulsars are located within supernova remnants, providing a dramatic confirmation of the Baade-Zwicky prediction.

1.6 Oscillations and orbits

Although the identification of pulsars with rotating neutron stars is secure, it is of considerable interest to recall the two other possible explanations for the source of the periodicity of the pulses that were discussed during the first few months after the discovery. The very precise periodicity might be due to the oscillation of a condensed star, or to a rapidly orbiting binary system. Both explanations were wide of the mark; nevertheless the discovery of the pulsars did stimulate new work on oscillations, involving a re-examination of the equation of state of condensed matter, while the binary theory soon found application in the X-ray pulsars and, later, in the relativistic dynamics of the binary pulsar discovered in 1974.

1.7 Oscillations

In 1966, shortly before the discovery, Melzer and Thorne showed that a white dwarf star could have a resonant periodicity of about 10 s, for radial oscillation in the fundamental mode. No means of driving the oscillation was proposed. The period was determined by a combination of gravity and elasticity, but it was not far from the simple result of calculation using gravity alone. Dimensional arguments show that the period is independent of radius and proportional to $(G\varrho)^{1/2}$ where ϱ is the density; for example, a white dwarf with density 10^7 g cm^{-3} would have a period of about 10 s if gravity alone provided the restoring force. Elasticity, which is in fact the dominant force, reduces the periodicity to the order of 1 s. No shorter period seems to be possible for a fundamental mode and higher order modes could not give such a simple pulse. The discovery of a pulsar with period 0.25 s among the first four therefore rules out the oscillating white dwarf as a possible origin.

Melzer and Thorne had also calculated the period of oscillation of neutron stars. Here the fundamental modes of radial oscillation had periods in the range 1 to 10 ms, and no possibility seemed to exist for lengthening the periods by the necessary two orders of magnitude.

The oscillation theories were soon completely overtaken by the discoveries of the short-period pulsars, Vela (89 ms) and Crab (33 ms), whose periods lay in the middle of the impossible gap between the theoretical oscillation periods of white dwarfs and neutron stars.

1.8 Planetary and binary orbits

Let us suppose that the pulsar period P is the orbital period of a planet, or satellite, in a circular orbit, radius R, round a much more massive condensed star with mass M_0 (in solar units). Then

$$R = 1.5 \times 10^3 \, M_0^{1/3} \, P^{2/3} \text{ km.} \tag{1.1}$$

It is therefore just possible for a satellite to orbit a white dwarf star of 1500 km radius with a period of 1 s, but the orbit would be grazing the surface. It would be more reasonable to consider a neutron star as the central object, when periods down to 1 ms would be possible. There are, however, two insuperable objections to the proposition that orbiting systems of this kind provide a model for pulsars.

The main difficulty concerns gravitational radiation, which is due to the varying quadrupole moment of any binary system. The energy loss through gravitational radiation would lead to a decrease in orbital period. A general formulation of the time scale τ of this change was given by

Ostriker (1968) for a binary system with masses M and εM, with angular velocity ω:

$$\frac{1}{\tau} = \frac{1}{\omega}\frac{d\omega}{dt} = \frac{96}{5}\frac{\varepsilon}{(1+\varepsilon)^{1/3}}\frac{(GM)^{5/3}}{c^5}\omega^{8/3}. \qquad (1.2)$$

For a satellite mass m where $\varepsilon = m/M$ is small, and $M = M_0$

$$\tau = 2.7 \times 10^5 \, \varepsilon^{-1} \text{ s}. \qquad (1.3)$$

The time scale was evidently far too short unless the satellite mass was very small. Pacini and Salpeter (1968) soon established that early observations of the stability of the period showed that m must be less than 3×10^{-8} solar masses.

Even the improbable hypothesis that such a small mass could be responsible for the radio pulses faced a second problem. The satellite would be orbiting in a very strong gravitational field, which would tend to disrupt it by tidal forces. Pacini and Salpeter showed that, even if it were made of high tensile steel, it could not withstand these forces unless it was smaller than about 20 m in diameter. An added problem would be that the satellite would be liable to melt or evaporate in the very high radiation field of a pulsar.

The same situation evidently obtained *a fortiori* for a binary system, for which a very rapid change in period would be expected. Planetary and binary systems were therefore eliminated as possible origins for the clock mechanism of pulsars. Gravitational radiation itself does, however, recur in the pulsar story; a pulsar was eventually found which is itself a member of a binary system with the short period of 7¾ hours, in which the orbital period decreases due to gravitational radiation at the rate of 30 ms per year (see Chapter 5).

1.9 Rotation and slowdown

The maximum angular velocity ω of a spinning star is determined by the centrifugal force on a mass at the equator. An estimate is easily obtained by assuming that the star is spherical with radius r; the centrifugal force is then balanced by gravity when

$$\omega^2 r = \frac{GM}{r^2}. \qquad (1.4)$$

This is, of course, the same condition as for a satellite orbit grazing the surface. If the star has uniform density ϱ, then the shortest possible rotational period P_{min} is roughly

$$P_{min} = (3\pi/G\varrho)^{1/2} \qquad (1.5)$$

1.9 Rotation and slowdown

A period of 1 s therefore requires the density to be greater than 10^8 g cm^{-3}, which is just within the density range of white dwarf stars. Neutron stars, on the other hand, can rotate with a period as small as 1 ms as demonstrated by the discovery of the 'millisecond' pulsar PSR 1937+21.

The limit on rotational angular velocity is somewhat more severe than in this simple argument, because the star will distort into an ellipsoid and tend to lose material in a disc-line extension of the equatorial region. The white dwarf theory was therefore already on the verge of impossibility for the first pulsars; the discovery of the short-period pulsars at once ruled it out completely.

The identification of pulsars with rotating neutron stars required the pulses to be interpreted as a 'lighthouse' effect, in which a beam of radiation is swept across the observer. This idea was supported by the observation by Radhakrishnan and Cooke (1969) that the plane of polarisation of radio waves from the Vela Pulsar swept rapidly in position angle during the pulse, which agreed with some simple models of beamed emission. The radio source must then be localised, and directional, as well as powerful. This led Gold (1968) to his seminal note in *Nature*, in which he suggested the identification with rotating neutron stars, the existence of a strong magnetic field, which drove a co-rotating magnetosphere, and the location of the radio source within the magnetosphere, probably close to the velocity-of-light cylinder. He also pointed out that rotational energy must be lost through magnetic dipole radiation, so that the rotation would be slowing down appreciably.

The early measurements of period on the first pulsar PSR 1919+21 showed that no change was occurring larger than one part in 10^7 per year. This limit was very close to the actual changes, which were measured a few years later, but the early null result could be used only to show that the stability of the period was in accord with the large angular momentum of a massive body in rapid rotation. Pacini (1968) showed that the limit on slowdown implied a magnetic field strength at the poles of a white dwarf less than 10^{12} gauss (10^8 tesla). He considered only magnetic dipole radiation in free space, which radiates away the rotational energy at a rate

$$\frac{dW}{dt} = -\frac{2\pi\omega^4 \mu_0}{3c^3} r^6 B_0^2 \sin^2 \alpha \qquad (1.6)$$

where B_0 is the polar field and α is the angle between the dipole axis and the rotation axis.

The slowdown of the Crab Pulsar was first measured by Richards and Comella (1969). From October 1968 to February 1969, the period lengthened uniformly by 36.48 ± 0.04 ns per day, i.e. by over 1 μs per month. The rate of change was consistent with the known age of the Crab Nebula, confirming the association of the pulsar with the supernova explosion observed in AD 1054. Furthermore, the rate of change could be applied to the neutron star theory, giving an energy output that was sufficient for the excitation of the continuing synchrotron radiation from the Crab Nebula. This coincidence was the final proof of the identification, as pointed out in Gold's second *Nature* letter (1969).

In retrospect, it is intriguing to consider what deduction might have been made from the measured variations of rotation period of the Vela Pulsar, if it had happened (as it nearly did) that those measurements had preceded those of the Crab Pulsar. The period of the Vela Pulsar was observed to be increasing slowly from November 1968 to February 1969, at the rate of 11 ns per day but, at the end of February, a discontinuous decrease in period occurred, amounting to 200 ns. The change occurred in less than a week (Radhakrishnan and Manchester 1969; Reichley and Downs 1969). By the time that this anomalous step was announced, the neutron star theory was already firmly established, and the decrease in period was regarded as an aberration rather than the typical behaviour. The step, or 'glitch', was interpreted solely on the basis of a change of moment of inertia, due to shrinkage or a 'starquake' (Chapter 6).

1.10 Optical pulses from the Crab Pulsar

The possibility that pulsars might emit pulses of light as well as radio was tested on the first pulsar, PSR 1919+21, as early as May 1968. In the excitement, some over-optimistic positive results were reported at first from both Kitt Peak and Lick Observatories but eventually every attempt was abandoned without any detection of optical pulsations or variation of any kind in several radio pulsars. Photometric equipment was, however, assembled for searches for periodic fluctuations in white dwarf stars and, on 24 November 1968, a recording of the centre of the Crab Nebula was made by Willstrop in Cambridge without prior knowledge of the discovery of the radio pulsar a few days earlier in the USA (Willstrop 1969). Although this recording was subsequently found to show the optical pulsations of the Crab Pulsar, it was stacked away with others for off-line computer analysis, and the discovery went instead to an enterprising team at the Steward Observatory in Arizona who were among three groups of observers fired with enthusiasm by the radio discovery of the Crab Pulsar.

The discovery of the optical pulses by Cocke, Disney and Taylor was published in a *Nature* letter (Cocke *et al.* 1969); less usually, the actual event of the discovery was recorded on a tape recorder, which was accidentally left running at the time. The excitement of the appearance of a pulse on a cathode ray tube, after a few minutes of integration, is well conveyed by the uninhibited (and unprintable) remarks of the observers. The discovery was made on 16 January 1969. Only three nights later the light pulses were observed by two other groups, at McDonald Observatory and Kitt Peak Observatory. Shortly afterwards, a new television technique was applied to the 120-inch reflector at Lick Observatory, and a stroboscopic photograph of the pulsar was obtained. This showed two contrasting exposures, made at pulse maximum and minimum (Fig. 1.2).

Subsequent observations have, of course, given very much more detail about the pulse timing, pulse shape, spectrum, and polarisation of these optical pulses; as might be expected, these are recorded in less dramatic form than the first paper by Cocke, Disney and Taylor, and their accidental historic tape recording.

1.11 X-ray pulses from the Crab Pulsar

The final link in the chain of discoveries about the Crab Pulsar was the extension of the spectrum into the X-ray region. The observations were necessarily made from above the Earth's atmosphere. In 1969 there was no X-ray telescope orbiting the Earth in a satellite, so that the only possibility lay in rocket flights. Astonishingly, two such rocket flights were successfully made, within a week of one another, and only three months after the discovery of optical pulses. The first was made by a team from the Naval Research Laboratory Washington (Fritz *et al.* 1969), and the second from the Massachusetts Institute of Technology (Bradt *et al.* 1969).

Both were completely successful, showing that the pulsed radiation extended to X-ray energies of several kilovolts; in fact, the total power radiated in the X-ray region was found to be at least 100 times that in visible light. The shape of the pulses was very nearly the same in X-rays as in light.

The Crab Nebula had been known and studied for several years as a source of X-rays. After the two rocket flights designed especially for the detection of periodic pulses had demonstrated the existence of the pulsar within the Nebula, the recordings of an earlier rocket flight were re-examined; they showed that the pulses had been recorded but not recognised. This flight was in March 1968 (Boldt *et al.* 1969). Even this pre-discovery recording turned out not to be the earliest, since a balloon-borne experiment in 1967 designed to measure the spectrum of the Crab Nebula up to X-ray energies of 20 keV was found to have recorded the periodic

Fig. 1.2. The Crab Pulsar. This pair of photographs was taken by a stroboscopic technique, showing the pulsar on (*above*) and off (*below*). (Lick Observatory, reproduced by kind permission of the Royal Astronomical Society.)

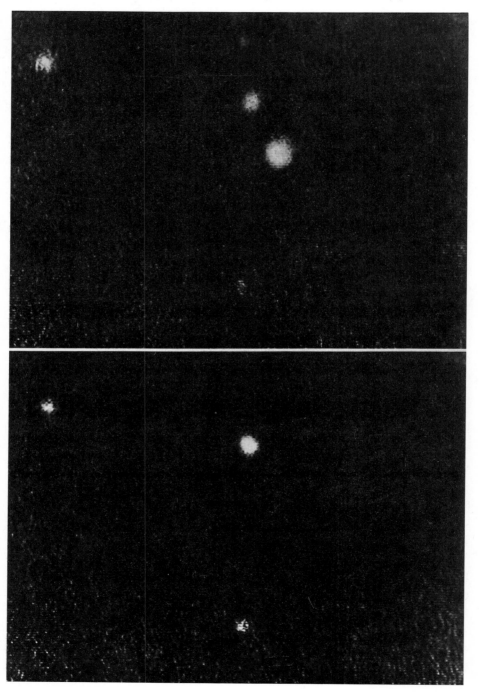

1.12 The development of pulsar research 15

'light curve' of the pulsar (Fishman, Harnden and Haymes 1969). There was sufficient accuracy in the periods obtainable from these two earlier experiments to show that the pulsar had been slowing down at the same average rate prior to the discovery as afterwards.

1.12 The development of pulsar research

The richness of the new field of research was demonstrated by a very rapid branching of the subject matter, even during 1968. There was, of course, from the start a division between theoretical and experimental papers, but the theories and observations arrived chaotically, without the dignified alternation of hypothesis and experimental test that is supposed traditionally to represent the progress of science.

When the dust had cleared from this initial explosion, the theorists were seen to be at work on several lines: the solid-state theory of the interior of neutron stars, the crystalline state of the surface, the magnetosphere, and the radiation mechanism which produced the pulses. The observers were ordering their work towards the determination of periods and positions, the organised search for more pulsars, the description of physical conditions in the pulse emitter, and the use of the pulses in exploring interstellar space. The results of these researches form the subject matter of this book.

The following chapters will show that considerable success has been achieved in the understanding of the interior of the neutron star, but that comparatively little is known about the magnetosphere and the mechanism of the radiation. The population of pulsars within the Galaxy is now known to fit reasonably well with the hypothesis that the neutron stars are created in supernova explosions. A complex story is unfolding, which accounts for the rapid rotation of the millisecond pulsars, and for the existence of a small number of binary pulsars. The exploration of interstellar space has led to a determination of the magnetic field in a large sector of the galactic plane and, through scintillation experiments, to a demonstration of very high velocities in the pulsars; the origin of these velocities appears to be related to the distribution of binary systems.

Apart from the pulsar discovery itself, the most outstanding success is the use of the binary pulsar PSR 1913+16 for the most comprehensive demonstration of relativistic dynamics, including the radiation of gravitational waves.

Over four hundred pulsars have now been discovered. Among these only two are extragalactic; these are PSR 0540−69 in the Large Magellanic Cloud, discovered as an X-ray source by Seward et al. (1984)

and later as an optical pulsar by Middleditch and Pennypacker (1985), and PSR 0042−73 in the Small Magellanic Cloud (Ables *et al.* 1987). Detection of the pulsars in more distant galaxies remains beyond our present means of observation.

2

The pulsars – an overview

Neutron stars, once a remote theoretical concept, now make their presence known in three ways: as the radio pulsars, as X-ray binaries, and probably as gamma-ray bursters. Of these three, by far the most numerous and the most informative are the pulsars. Their radio emission represents only a small part of the total energy generated by their rotational slowdown, but it is nevertheless a remarkable indicator of many physical properties. It is even possible to explore conditions deep in the interiors of the stars. Over 400 pulsars have been discovered, a sufficient number for significant studies of their population, their life history, and their distribution through the Galaxy. This chapter gives a brief survey of their properties and their significance for astrophysics, serving as an introduction to the more detailed chapters that follow.

2.1 Periods

The pulsating radio signals, which gave pulsars their name, are derived from a lighthouse beam, on a rotating neutron star. The beam is locked to the solid crust of the star by a very strong magnetic field, so that the pulses allow the rotation to be followed precisely. Most pulsars have periods between ¼ and 2 seconds; the longest period known is 4 seconds. All periods are lengthening as they slowly lose their kinetic energy of rotation; for the majority of pulsars the rotation slows down on a timescale of 10^6 to 10^8 years. The distribution of periods and slowdown rates suggests that most pulsars start their lives with periods below 100 milliseconds, follow similar evolutionary paths, and cease to radiate after a few million years. This interpretation leads to a birthrate of order 1 per 50 years, consistent with an origin in the violent collapse of massive stars which we see as supernovae.

A small separate population of pulsars, mostly with very short periods, have a much smaller rate of slowdown, due to much smaller magnetic fields. These 'millisecond' pulsars are regarded as neutron stars that have passed the normal age span of activity; their magnetic fields have decayed with time, but they have been rejuvenated by a spin-up process involving a binary partner. This spin-up process itself can be observed directly in the X-ray binaries; here the accretion of matter from a companion provides both the thermal energy for the X-ray emission and the angular momentum to increase the rotation speed.

In contrast to the majority of stars, most pulsars are not in binary systems. Only about two per cent are now known in binaries; of these most are characteristic of the millisecond pulsars, and their rejuvenation may be due to the binary system in which they are now observed. There is, however, a strong indication that many members of the normal population were originally in binary systems, which were disrupted at the time of their birth in supernova explosions.

2.2 Galactic population

Surveys of pulsars, designed to discover how they are distributed through the Galaxy, are subject to the obvious limitations of sensitivity: weak pulsars can only be seen out to short distances, and may therefore be poorly sampled, while the majority may be difficult to see as far away as the galactic centre. In addition there are severe limitations due to the effects of the interstellar medium on radio propagation; this is particularly important close to the galactic plane and towards the galactic centre. Despite these problems and the as yet incomplete coverage of the sky, a statistically sound result has emerged. There is a total population of between 10^5 and 10^6 active pulsars in the Galaxy. They are concentrated in the plane of the Galaxy within a layer about 1 kiloparsec thick and within a radial distance of about 10 kiloparsecs from the centre. Measurements of their motion show that they have high velocities, presumably originating in their violent births, and are mostly moving away from the plane at a rate of order 100 km s^{-1}, so that the distribution is consistent with an origin within 100 parsecs of the plane; this is to be expected if they represent the end product of the evolution of massive stars.

The millisecond pulsars represented a smaller population of old pulsars. Their rate of 're-birth' in the spin-up process is much lower than the birthrate of the normal pulsars, and they are only observable as a considerable population because of their long life.

2.3 The structure of neutron stars

Theoretical models of neutron stars show that the allowable range of mass is between about 0.2 and 2.0 solar masses ($M_\odot$); a smaller mass

2.3 The structure of neutron stars

provides insufficient gravitation to hold the star together in its condensed state, and a larger mass would lead to further collapse to a black hole. The larger the mass, the more concentrated is the star; the moment of inertia therefore tends to be nearly constant throughout the allowable range of mass. The diameter of a typical neutron star with mass $1.4\,M_\odot$ is 20–30 km, depending on the equation of state of the interior; the central density is between 10^{14} and 10^{15} gm cm^{-3}.

The equation of state, which is deduced from high-energy particle interactions, is uncertain for the highest densities, i.e. near the centre of the star. Stiffer equations of state, for which there is most theoretical support, are those in which the neutron fluid is less compressible; they lead to a larger radius, a lower central density, a thicker solid crust, and a moment of inertia of order 3×10^{44} g cm^{-3}. (For comparison, the moment of inertia of the Earth is 7×10^{33} g cm^{-3}.) Detailed calculations may be found for example in Arnett and Bowers (1977).

The structure of a typical model neutron star with mass $1.4\,M_\odot$ is shown in Fig. 2.1. Between the crust and the centre the density covers a range of about nine orders of magnitude, from 10^6 to 10^{15} g cm^{-3}. The crust is a very rigid and strong crystalline lattice, primarily of iron nuclei. At higher densities it becomes energetically favourable for electrons to penetrate the

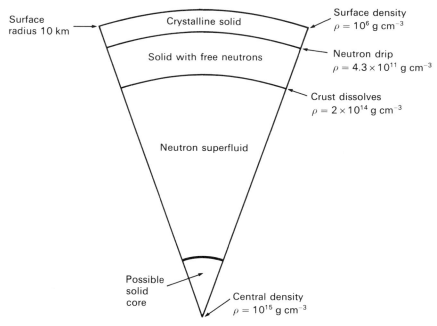

Fig. 2.1. Typical cross-section of a neutron star. Neutron stars with low mass may not have a solid core and their outer layers will be more extended.

nuclei and combine with protons, forming nuclei with unusually high numbers of neutrons, including species unknown in the laboratory such as ^{118}Kr, a remarkable nucleus with eighty-two neutrons and thirty-six protons (Baym *et al.* 1971).

Up to a density of 4×10^{11} g cm^{-3}, there are almost no neutrons outside the nuclei. Above this density, referred to as the neutron drip point, the most massive nuclei become unstable and the nuclei are embedded in a neutron fluid. The central core consists only of a neutron fluid, containing a small proportion of electrons and protons. There remains a possibility (*cf.* Irvine 1978) that a solid core may exist for densities in excess of 3×10^{14} g cm^{-3}. The neutron fluid is a superfluid, with no viscosity; this is important in the explanation of irregularities of pulsar timing (Chapter 6).

2.4 The magnetic fields of neutron stars

The pulsars are very strongly magnetised neutron stars. Their dipole field strengths are consistent with the collapse of a normal star with a field of order 100 gauss (10^{-2} tesla), the flux being conserved in the collapsing stellar material. Polar field strengths reaching 10^{12} gauss occur in young pulsars; in old pulsars the field may fall to 10^{10} gauss, while in the 'millisecond' pulsars it may be as low as 10^{8} gauss. Despite the intensity of the magnetic field, it has very little effect on the structure of the star. The energy density is very high, corresponding to an equivalent matter density of order 1 kg cm^{-3}, but the only effect is a modification of the crystal structure near the surface (Ruderman 1974). Outside the star, however, the magnetic field completely dominates all physical processes, outweighing gravitation by a very large factor. The ratio

$$\frac{GMm}{r^2} \bigg/ \frac{e\Omega rB}{c}$$

between the gravitational and induced electrostatic forces on an electron near the surface of the Crab Pulsar is of order 10^{-12}.

The external magnetic field plays a crucial part in several observable characteristics of pulsars. The dipole is substantially misaligned with the rotation axis, and the dipole therefore generates an electromagnetic wave at the rotation frequency. This accounts for the main loss of rotational energy and the observed slowdown. For a young pulsar the outward energy flow is sufficient to provide high-energy particles in a surrounding nebula; the rotational energy of the Crab Pulsar is the source of energy for the synchrotron radiation of the surrounding Crab Nebula.

A local electric field is induced by the rotating magnetic field and this is

2.4 The magnetic fields of neutron stars

an overwhelming influence in a region from the pulsar surface out to a radial distance $r_c = c/\omega$, i.e. the distance where a co-rotating extension of the pulsar, with angular velocity ω, would have a speed equal to the velocity of light c. This radial distance defines the 'velocity of light cylinder'. Inside this cylinder there is an ionised 'magnetosphere' of high-energy plasma, co-rotating out to a distance approaching r_c. It is within this magnetosphere that the beam of radiation originates (Fig. 2.2). Goldreich and Julian (1969) analysed the fields and charge densities expected to be built up in the magnetosphere for the simplest case in which the magnetic field is aligned with the rotation axis. The magnetosphere is highly conducting along, but not perpendicular to, the magnetic field lines. This condition in the magnetosphere is similar to the high conductivity of the stellar interior, where there can be no net electric field. The magneto-

Fig. 2.2. The essential features of a pulsar magnetosphere. Within a radial distance $r_c = c/\omega$ of the rotation axis there is a charge-separated, co-rotating magnetosphere. The magnetic field lines which touch the velocity-of-light cylinder at radius r_c define the edge of the polar caps. Radio emitting regions in the polar caps are shown cross-hatched.

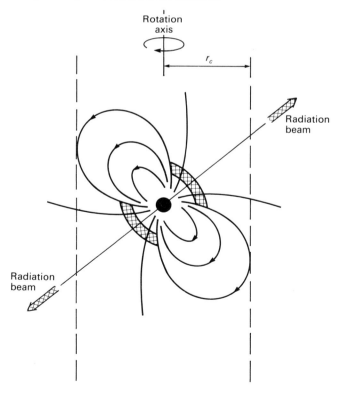

sphere then seems to be an extension of the solid interior; in both regions the induced electric field is cancelled by a static field, so that

$$\mathbf{E} + \frac{1}{c}(\mathbf{\Omega} \times \mathbf{r}) \times \mathbf{B} = 0$$

In this fully conducting situation there must be a charge density equal to $(1/4\pi)\mathrm{div}\,\mathbf{E}$; a simple analysis then shows that the difference in numbers of positive and negative charges is given by

$$n_- - n_+ = \frac{\mathbf{\Omega} \cdot \mathbf{B}}{2\pi ec}$$

As a useful guide, the particle density is given approximately by

$$n_- - n_+ = 7 \times 10^{-2}\, B_z P^{-1}\, \mathrm{cm}^{-3}$$

where B_z is the axial component of the field in gauss, and P is the period in seconds. Opposite signs of net charge are found in the equatorial and the polar regions.

A more detailed analysis of the aligned rotator can be found in Fitzpatrick and Mestel (1988). The general case of the non-aligned rotator has so far defied all attempts at full analysis, although it appears that many of the characteristics of the magnetosphere must be similar to those of the aligned rotator (Mestel 1971).

The magnetic field also exerts a powerful dynamical influence on the interior of a pulsar. The neutron fluid contains a small proportion of electrons and protons; this charged component couples the fluid strongly with the magnetic field. The part of the fluid core which penetrates the inner solid crust is not, however, completely coupled, so that the rotation of the neutron star is comprised of a solid-body rotation and a fluid rotation, the coupling between these components being variable. The coupling can in fact change discontinuously; the rotation rate then shows a step, known as a 'glitch'. These glitches prove to be valuable indicators of conditions inside neutron stars.

X-ray binaries also contribute to our knowledge of the magnetic field of neutron stars. The matter accreted from the binary companion onto the neutron star is constrained to fall on restricted regions at the magnetic poles; these heated regions are responsible for the X-ray emission. A cyclotron resonance can occur in the ionised accreting material, giving a spectral line in the X-ray domain which directly measures the strength of the magnetic field.

2.5 Radio pulse characteristics

For most pulsars, the radio pulse occurs during a small fraction of the period, corresponding to between 5° and 20° of angular rotation. The pulse is formed by a beam of radiation that sweeps across the observer's line of sight, as in a lighthouse. The beam is radiated by high energy particles constrained to move along the field lines over a magnetic pole, so that the beam is rigidly attached to the solid surface of the neutron star. The radio beam originates in regions some tens of stellar radii above the surface; the lower the radio frequency, the further from the surface is the emitting region. Optical and other high energy radiation, only observed from the young pulsars, originates in a separate region, also tied to the magnetic field lines, but closer to the velocity-of-light cylinder.

The individual radio pulses from a particular pulsar are very variable, as though the lamp in a lighthouse were flickering and the beam direction were varying erratically through a small range. The superimposition of some hundreds of recorded pulses nevertheless gives an integrated pulse profile, which is stable and characteristic for each individual pulsar.

Some integrated pulse profiles extend over a large proportion or even the whole of the pulsar rotation period. The interpretation of this apparently very wide lighthouse beam may be a near alignment of the magnetic and spin axes. Other integrated profiles show two distinct components, representing two beams radiated from the two opposite poles of a dipole field oriented nearly perpendicular to the rotation axis.

The individual radio pulses are often very highly polarised. When the integrated pulse profile is formed by adding many pulses, and the polarised components are also added suitably, a high average polarisation may be found. This often shows a remarkably simple behaviour: the position angle of the linear component swings monotonically through the integrated pulse by an angle up to 180°. A circular component is also seen, usually close to the centre of the pulse; this may be of one hand throughout, or it may reverse polarity. If it reverses, it usually does so only once.

This simple pattern of polarisation indicates that the origin of the radio beam is above a magnetic pole; the swing of position angle is then related to radio emitting regions whose radiation is polarised along the direction of the field lines. This interpretation is supported by the observation that the pattern of polarisation in a given pulsar is very similar over a wide range of radio frequencies; nevertheless it is difficult to interpret the amplitude data in detail, and it is evident that the radio beams do not have a simple geometrical configuration.

The intensities of the emitted radio emission are extremely high, so that they cannot be due to thermal emission or to incoherent synchrotron emission. The radio pulses are also extremely variable in intensity on several time scales. Many pulsars have subpulses containing narrow spikes; some have quasi-periodic sequences, referred to as 'periodic microstructure'. This fine structure is usually regarded as a modulation of the radiation process, rather than a beam structure observed as the pulsar rotates.

Subpulses may appear independently at each pulse, or they may appear in sequences; in these sequences they may appear progressively earlier or later in relation to the integrated pulse profile. This 'pulse drifting' is characteristic of individual pulsars; some have several different characteristic rates at which drifting may occur.

There may also be more than one integrated pulse profile associated with an individual pulsar. For a period up to several hours one profile is observed; a switch to a different 'mode' then occurs without warning. The overall polarisation characteristics do not change, however; the interpretation is that the excitation of the emission process has switched to a different distribution across the emitting region. There are also 'nulls', occasions when the emission stops completely for a number of rotation periods.

2.6 Optical and other high energy emission

The radio emission from pulsars represents a very small fraction of the energy available from the observed slowing down of rotation. The youngest pulsar, the Crab Pulsar, emits detectable pulses over the whole spectrum from infrared to high energy gamma-rays; most of the radiated energy is in the X-ray region. This high-energy radiation is in the form of a rotating lighthouse beam, like the radio emission of older pulsars: unlike the usual radio emission, it originates high in the magnetosphere, close to the velocity-of-light cylinder. In the emitting region there are charged particles, probably electrons and positrons, with energies up to 10^{15} eV, constrained to follow magnetic field lines in a restricted region. The radiated beam from this region has a continuous spectrum and a similar shape from infrared to high energy gamma-rays. There is also radio emission directly associated with it, giving a total range of more than 50 octaves of the electromagnetic spectrum.

High-energy radiation appears to be characteristic of the youngest pulsars. The Vela Pulsar is also a strong gamma-ray source; its high-energy radiation is in a double pulse, with separation up to 140°, although its radio emission is a single unrelated beam typical of the older pulsars.

The accreting neutron stars of the X-ray binaries do not display pulses corresponding either to the radio pulsars or to the high-energy young pulsars. Their radiation is thermal, originating in very hot in-falling material over the magnetic poles.

2.7 The interstellar medium

Radio waves propagating through interstellar space encounter a tenuous ionised gas, whose refractive index differs appreciably from unity. The effect is a delay in the arrival of a radio pulse, depending on the observing frequency and on the total electron content of the propagation path. The delay can be measured by comparing pulse arrival times at different radio frequencies, giving a 'dispersion measure'.

The interstellar medium is turbulent, with irregular structure on many scales. Radio waves do not therefore propagate on a single ray path with a plane wavefront. Multipath propagation, through refraction in the electron clouds, spreads the arrival time of an individual pulse; this 'pulse smearing' is especially noticeable in the more distant pulsars in the galactic plane. A smaller amount of scattering can still produce major effects through interference between waves propagating along paths whose lengths differ by only a few wavelengths; this is the phenomenon of 'scintillation'. The largest scales of turbulence produce slow variations of intensity, while smaller scales result in a more rapid fading. The timescales of these two types of fading depend on the radio frequency of observation; at low radio frequencies they may be some years and some tens of minutes respectively.

Faraday rotation of the plane of linear polarisation also occurs on the line of sight to all pulsars, due to the combined effect of electrons and the interstellar magnetic field. A combined measurement of dispersion measure and rotation measure, both obtained from multi-frequency observations, gives the component of the magnetic field along the line of sight. Such measurements reveal interstellar magnetic fields of a few microgauss. Since these measurements can be made over large distances, they are important in mapping the magnetic field of a large part of the Galaxy.

3

Searches and surveys

3.1 The problem of sensitivity

The Cambridge discovery of pulsars presented a challenge to other radio astronomers, with the question, 'Why had they not been detected before, in the many and varied surveys of the radio sky?' In fact, some existing sky survey recordings already showed unrecognised pulsar signals in which the recording technique had obscured the pulses. It was soon found that for all but the strongest pulsars the signals are too weak to be detected except by adding a long sequence of pulses, superposed at the periodicity of the pulsar. Evidently, only the largest radio telescopes can be used for pulsar surveys and only rarely can individual pulses be detected. Only the first pulsars discovered at Cambridge and in Australia, and later the Crab Pulsar, were discovered through simple recordings of individual pulses. In all subsequent searches and surveys the sensitivity has been increased by using a long integration time, achieved by the superposition of many pulses at a precise periodicity. The problem is, of course, that in a search the period is not known, and the superposition must be repeated over the whole range of possible periods. Moreover, the pulse width and the dispersion measure are also unknown and we have to search in these parameters as well. This process, which is analogous to Fourier analysis, involves a large-scale digital computation.

Most surveys of the sky in which pulsars have been discovered have used radio telescopes operating at wavelengths in the region of 70 centimetres (400 MHz). Pulsars can be observed at wavelengths from about 15 m to 3 cm (frequencies 20 MHz to 10 GHz); the choice of wavelength to give the best sensitivity involves a balance between several factors, and particularly the relation between the radio spectrum of pulsars and the galactic background against which they must be detected. The bandwidth of the receiver is critically important; as we shall see, the sensitivity increases with the bandwidth, but too large a bandwidth, particularly at low radio frequencies, can mask the periodic signal through dispersion in pulse arrival time. Modern techniques retain the high sensitivity resulting from large bandwidth by using an array of narrow-band receivers and conducting a search in dispersion measure.

The background noise level in a typical large radio telescope operating at 1 metre wavelength corresponds to a flux density of order 100 Jy (1 Jy = 1 Jansky = 10^{-26} W m^{-2} Hz^{-1}). Observable pulsar flux densities are of order 10^{-4} smaller. Reaching such a small fraction of the receiver noise level is commonly achieved in radio astronomical observations by using a wide receiver bandwidth B and averaging the signal over an integration time τ. The smallest detectable signal is then a fraction $(B\tau)^{-1/2}$ of the total noise level. Typically, a receiver might have a bandwidth of 1 MHz, and it may be required to detect a single pulse 10 milliseconds long; the sensitivity would then be 10^{-2} of the input noise, or about 1 Jy in a large radio telescope. Single pulses are rarely as strong as 1 Jy, and weaker pulses would be lost in noise. However a series of 10 000 pulses, suitably added together, would provide a further factor of 10^{-2}, giving a more useful sensitivity of 10^{-2} Jy within the averaged pulse; the mean flux density of a detectable pulsar will, of course, be lower than this, depending on the ratio of pulse width to period.

3.2 Frequency dispersion in pulse arrival time

Although the basic characteristic of the pulsar signal, which facilitates its recognition, is the precise periodicity, a second characteristic is the frequency dispersion in arrival time due to the ionised interstellar medium. This may assist the recognition of a pulsar signal against locally generated impulsive radio interference, but it also plays an important part in restricting the range of a pulsar search.

Radio pulses travel in the ionised interstellar medium at the group velocity v_g, which for small electron densities is related to the free space velocity c by

$$v_g = c \left[1 - \frac{n_e r_o \lambda^2}{2\pi} \right]$$

where $r_o = e^2/mc^2$ is the classical radius of the electron and n_e is the electron number density. Hence the delay t in travel time over distance L, compared with free space, is

$$t = \frac{n_e r_o c v^{-2}}{2\pi} L$$
$$= 1.345 \times 10^{-3} \, v^{-2} \, n_e L \text{ s}.$$

Customary units in astrophysics are parsecs (3×10^{18} cm) for distance and cm^{-3} for density: the product $n_e L$, which measures the total electron content between the pulsar and the observer, is known as Dispersion Measure, DM, with units pc cm^{-3}. Observers usually quote radio frequencies in megahertz, so that the delay becomes

$$t = 4.15 \times 10^3 \, \text{DM} \, \nu_{\text{MHz}}^{-2} \, \text{s}.$$

The frequency dependence of this delay has a very important effect on observations of radio pulses. A short broad-band pulse will arrive earlier at higher frequencies, traversing the radio spectrum at a rate

$$\dot{\nu} = \frac{\nu_{\text{MHz}}^3}{8.3 \times 10^3 \, \text{DM}} \, \text{MHz s}^{-1} \, ;$$

correspondingly, a receiver with bandwidth B_r (MHz) will stretch out a short pulse to a length

$$\Delta t = 8.3 \times 10^3 \, \text{DM} \, \nu_{\text{MHz}}^{-3} \, B_r \, \text{s}.$$

As a useful guide, $\Delta t = (202/\nu_{\text{MHz}})^3$ DM milliseconds per megahertz bandwidth.

If a pulsar with high dispersion measure is observed with a receiver with a wide bandwidth, its pulse is stretched and the peak intensity is reduced. The lost sensitivity may, however, be recovered by dividing a wide receiver bandwidth into separate bands and using separate receivers on each as in Fig. 3.1. The output of these separate receivers can then be added with appropriate delays so that the pulse components are properly superposed. This process of 'de-dispersion' is shown in Fig. 3.2.

The technique first used for 'de-dispersion' involved a mechanically-driven sequential sampling of the separate receiver outputs (Large and Vaughan 1971). Digital techniques are now universally in use, either setting the individual delays to match the known DM of a given pulsar, or using a series of different delays in an off-line search through recorded data. Each set of delays then allows the detection of pulsars in a definite range of DM.

3.3 Search techniques

The original Cambridge discovery depended on an unusual combination of radio telescope parameters, viz long wavelength, short integration time, very large collecting area, and regularly repeated observations. In the same way any new search for pulsars will be sensitive to pulsars with a range of characteristics depending on several parameters of the search technique, and in particular on radio frequency, bandwidth,

3.3 Search techniques

Fig. 3.1. Frequency dispersion in pulse arrival time for PSR 1641−45, recorded in 64 adjacent frequency channels, each 5 MHz wide, centred on 1540 MHz.

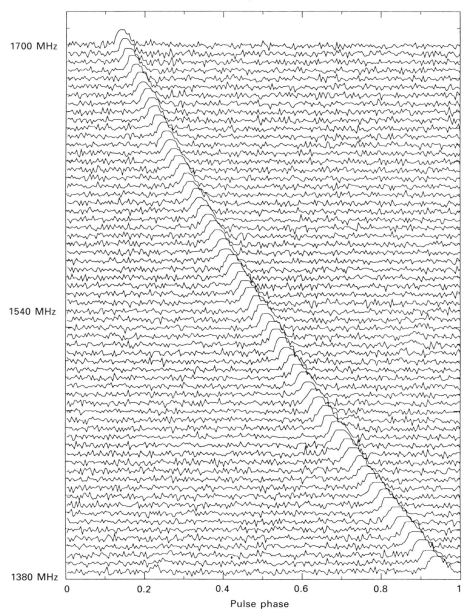

digital sampling rate and integration time. We need to examine the selection effects imposed by any particular choice of these parameters.

The first surveys, at Cambridge and at Molonglo, essentially scanned the northern and southern skies respectively for pulsars strong enough to be detected by the appearance of single pulses on pen chart recordings. The Molonglo radio telescope proved to be particularly suitable for pulsar search, and twenty-eight pulsars were found by inspection of survey recordings (Vaughan and Large 1969). The survey was extended to higher values of DM by the use of 'de-dispersion' in twenty separate receiver channels, but only two more pulsars were discovered by this technique, presumably because the higher values of DM corresponded to the more distant and therefore weaker pulsars.

In some early searches, the dispersion effect itself was used as a means of recognition of individual pulses against a background of radio noise and impulsive interference. The discovery of the Crab Pulsar by Staelin and Reifenstein was made using a receiver with fifty separate channels covering 110 MHz to 115 MHz. The Crab Pulsar is unusual in its production of occasional very strong individual pulses (the 'Giant' pulses: see Chapter 13). Some of these showed on the recordings, with a dispersion of over a second in the total bandwidth. The dispersion measure of the Crab Pulsar (DM = 57) is in fact so large that, even within a single channel 100 kHz

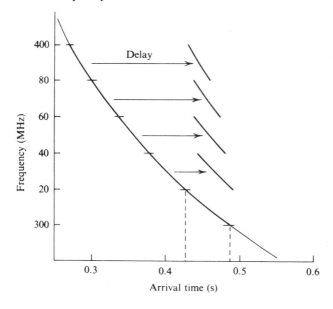

Fig. 3.2. De-dispersion, achieved by sequential sampling and delay of adjacent frequency channels.

3.4 Search for periodic pulses

wide, a regular pulsation would be smeared out over the 33-ms period; the discovery depended on the individual giant pulses and not on the periodicity, which was found later in observations at higher radio frequencies.

A more extensive search for dispersed single pulses made at Jodrell Bank used two receiver channels separated by 2 MHz at 408 MHz. Three pulsars were discovered in this search (Davies and Large 1970).

These searches seem to have exhausted the possibilities for finding pulsars by the detection of single pulses. We now turn to the more productive searches which depend on their characteristic precise periodicity.

3.4 Search for periodic pulses

Hidden within the apparently random noise at the detector output of a radio telescope pointing at an arbitrary region of the sky there may be a low-level, precisely periodic signal of unknown pulse length and unknown period. The detected signal can be sampled and digitised at an arbitrary rate, and an arbitrary length of record can be searched by computer for the periodic signal. The principle of the search may be to look either directly for a train of regularly spaced pulses (a periodogram analysis) or for their spectrum within the Fourier transform of the data stream. The two approaches are closely related, but the Fourier transform method is used in all modern searches because it is more economical of computer time.

The periodogram analysis is conceptually easy to follow. The data stream of N equispaced samples is folded at a series of N different values of period P, each operation providing a profile that may contain a pulsar signal. Each profile then has to be searched for pulses of different widths and phases, by scanning with a window function of variable width, since the width W may be anywhere between 1% and 50% of the period. The sensitivity depends on the pulse width; a narrow pulse, with width W, can be detected with a signal-to-noise ratio increased by a factor of about $(P/2W)^{1/2}$ over a sinusoidal waveform. This dependency of sensitivity on pulse width is fundamental in all search processes.

The Fourier analysis approach involves taking the fast Fourier transform (FFT) of the time series and inspecting the resultant spectrum for signals together with any associated harmonics. Fig. 3.3 shows a sketch of the Fourier relationship between the time and frequency domains for a pulse train. The amplitudes and phases of the individual harmonics are determined by the average pulse profile in the time domain. For a nearly sinusoidal pulse there will be a large fundamental spectral feature with small harmonics, while for a narrow pulse width W there will be approximately $P/2W$ harmonics with amplitude comparable to the fundamental. Individually, these components may not be distinguishable from noise, and they must be combined in some way to maximise the detectability of the signal. This is achieved in practice by an incoherent addition of the harmonics, i.e. by adding their amplitudes only, in the process illustrated in Fig. 3.4.

Here we see a spectrum containing several harmonics, and the same spectrum expanded so that the fundamental coincides with its second harmonic. For a large signal, the sum of these two components is almost double the fundamental, while the noise increases by only $2^{1/2}$, giving a net gain in signal-to-noise ratio. In a spectrum with many elements, say 10^5, this may greatly increase the significance of a suspected spectrum component. For example, a fundamental might appear with a signal-to-noise ratio of 4, a value occurring frequently by chance; in the combined spectrum this would be increased to 5.7, a value unlikely to occur by chance.

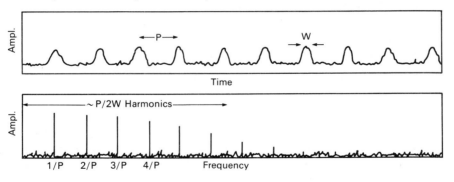

Fig. 3.3. The Fourier relationship between the time and frequency domains for a pulse train with period P and pulse width W.

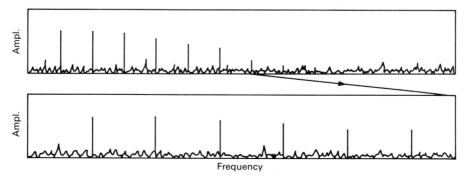

Fig. 3.4. The first half of the spectrum is expanded and added to the unexpanded spectrum so that each fundamental is added to its second harmonic.

3.5 The pulsar surveys

The process of expansion and addition may, of course, be repeated several times, resulting in the summation of more and more harmonics and providing good sensitivity to narrower pulses.

The search in the domain of dispersion measure is usually carried out in one of two ways. The most obvious is to repeat the spectrum analysis for a series of values of dispersion measure, i.e. by adding the outputs of the separate receiver channels with a series of different sets of delays. Alternatively, the data are treated as a two-dimensional array arranged as sample time against receiver channel. A two-dimensional FFT is then performed on the array, giving a spectrum as shown in Fig. 3.5. The pulses are now detected as a series of harmonically related features lying along a diagonal passing through the origin, with slope determined by the dispersion measure. The search proceeds by extracting diagonal segments from the array and seeking fundamentals and related harmonics as in the one-dimensional analysis.

Modern computers can perform large FFTs with great efficiency, while the other processes can be carried out in an even shorter time. The full analysis can therefore be used on very long data strings, which are essential for searches for pulsars with short periods and low intensity.

3.5 The pulsar surveys

The essential statistical information on the distribution of pulsars in location, periodicity and luminosity (Chapter 10) is derived from the

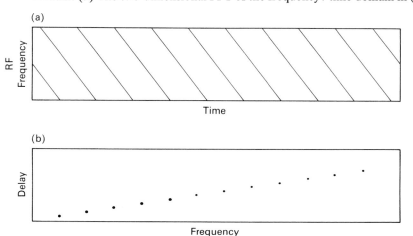

Fig. 3.5. (a) The appearance of dispersed pulses in the RF frequency / time domain. (b) The two-dimensional FFT of the frequency / time domain in (a).

following main surveys, mostly made at a radio frequency of about 400 MHz (wavelength 70 cm) (see also Manchester 1987).

(i) Large and Vaughan (1971). Thirty-one pulsars were discovered, using the east–west arm of the Molonglo Cross telescope. Three steradians of the southern sky were searched for single pulses having dispersion measure up to 400 cm^{-3} pc.

(ii) Davies, Lyne and Seiradakis (1972, 1973). Thirty-nine pulsars were discovered, using the 250-ft Lovell telescope at Jodrell Bank. The survey was sensitive to periods between 0.08 and 4.0 seconds, and covered 1 steradian along the galactic plane. Its limitation in dispersion measure is indicated in Fig. 3.6.

(iii) Hulse and Taylor (1974, 1975). Forty pulsars were discovered, using the 1000-ft Arecibo radio telescope. This survey covered a small solid angle, 0.05 steradian, in the galactic plane. The limiting sensitivity was 1.5 mJy, about a factor of 10 better than previous surveys. The search

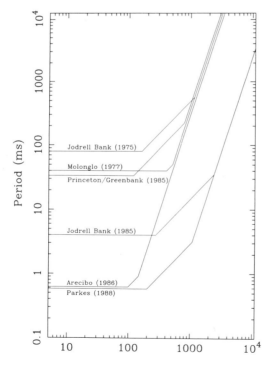

Fig. 3.6. The range of period and dispersion measure covered by various surveys. Surveys have little sensitivity below the indicated lines.

covered periods from 0.03 to 3.9 seconds, and extended to DM = 1280 cm^{-3} pc.

(iv) Manchester *et al.* (1978). The second survey at the Molonglo radio telescope in Australia followed the single-pulse surveys described above. In the second survey 155 pulsars were discovered, the largest total in any survey. The whole sky south of declination +22° was covered, complete to a sensitivity of 15 mJy for periods greater than 40 ms.

(v) Stokes *et al.* (1985) and Dewey *et al.* (1985). These observations constituted part of the Greenbank Survey of the northern hemisphere and utilised the 300-ft transit paraboloid at Greenbank, West Virginia. They provided a complement to the second Molonglo survey of the southern sky and had a similar sensitivity. Fifty-four new pulsars were discovered. Although these searches had improved sensitivity for shorter periods, they did not reveal the presence of many hitherto undetected short period, 'young' objects like the Crab Nebula pulsar.

(vi) Clifton and Lyne (1986). This survey, using the 250-ft Lovell telescope at Jodrell Bank was sensitive to pulsars with short periods ($P \geqslant 0.004$ seconds) and large dispersion measures (DM $\geqslant 35000P$ cm^{-3} pc). This was achieved by using the comparatively high frequency of 1400 MHz and eight contiguous channels, each 5 MHz wide. Forty pulsars were discovered, most with large dispersion measures. The survey covered the galactic plane in the region $-4° \geqslant l \geqslant 105°$, $|b| < 1°$.

3.6 Selection effects – the limits of the surveys

Apart from the hope of discovering interesting individual pulsars, the main object of a survey is to measure the population of pulsars as a function of parameters such as period, flux density, spectrum and location within the Galaxy. The limits of sensitivity within each search must therefore be carefully delineated. For example, a high sensitivity can only be achieved by using a long integration time, while the shortest periodicity requires a rapid sampling rate; the two together determine the total number of samples in an integration, which is limited by available possibilities of data-processing, since the observations commonly require million-point Fourier transforms as part of the processing.

The useful sampling rate is limited by the range of dispersion measure which is to be covered. Even when the receiver bandwidth is divided into separately detected bands to avoid the effects of dispersion, a procedure that considerably increases the amount of data recording and computation, there is a fundamental limit due to interstellar scattering (Chapter 17), which stretches pulses so that the periodicity may be smoothed out.

The 1986 Jodrell Bank survey was designed with this limitation particularly in mind. This survey was intended to search for pulsars with very short periods located close to the galactic plane, where scattering is most important. Fig. 3.6 shows the limitations of this and other surveys in respect of period and dispersion measure, using the relation found empirically by Slee *et al.* (1980) between DM and pulse lengthening τ_{scatt}:

$$\tau_{scatt} \approx \left(\frac{DM}{1000}\right)^{3.5} \left(\frac{400}{\nu_{MHz}}\right)^4 \text{ milliseconds.}$$

Although the main population of pulsars probably lies within the area at the upper left of this diagram, the limitations of the earlier periodicity searches are obvious. For example, the Crab Pulsar ($P = 33$ ms, DM $= 57$) would not have been found by its periodicity alone, while the 'millisecond pulsar' PSR 1937+21 has such a short period (1.6 milliseconds) that it would not be found even in the 1986 survey.

In searches for longer period pulsars the sampling rate can be low and a long integration time then becomes possible. For example, the first pulsar discovered in the Small Magellanic Cloud (PSR 0042−73, Ables *et al.* 1987) has a mean flux density of only 1.3 mJy; the long period of 0.926 seconds allowed a total integration time of 87 minutes to be used in the search.

3.7 Searches for millisecond pulsars

The most interesting pulsars, which must account for as much published work as all the others put together, are the Crab Pulsar, the millisecond pulsars, and the binary pulsar PSR 1913+16. Of these, only the third has a signal large enough and a period long enough to have been discovered in any of the main surveys up to mid-1982; even this pulsar might have been rejected in a large-scale survey because of its varying period arising from its orbital motion. The main reason for the relative inadequacy of these surveys was the lack of rapid data recording and computing resources. The problem is particularly acute for millisecond pulsars because, for a given dispersion measure, the amount of data to be recorded and processed depends on the inverse square of the period. For instance, a doubling of the sampling rate to detect shorter period pulsars has to be matched by a doubling of the number of frequency channels to be sampled within a total receiver bandwidth, so that dispersion broadening can be halved and remain the same fraction of the period. Thus, for example, the move from the searches of the 1970s with sample rates of typically 10 ms to the millisecond pulsar surveys of the 1980s with sample rates of 0.3 ms required increases of about three orders of magnitude in data storage and computer capacity.

3.7 Searches for millisecond pulsars

Large-scale searches for millisecond pulsars were impossible before these techniques and resources became available. Instead, the searches were confined to a small number of interesting objects suspected of being pulsars and to a number of interesting locations, such as supernova remnants and globular clusters, where there were reasons to suspect the presence of millisecond pulsars. There have been several such 'targetted' searches (Backer 1987). Such searches can reasonably occupy more time than the search of an individual area in a major survey, and they can therefore extend to a lower level of sensitivity and, most important, to shorter periods. Even so, if there is a long list of candidates, the total time of survey and analysis is not inconsiderable, and it is important to pick out a good list by setting down the most likely characteristics for pulsar candidates.

Pulsar candidates are expected to have small angular diameters, steep radio spectra, and high linear polarisation. All three characteristics were found in the radio source 4C21.53, subsequently found to contain the millisecond pulsar PSR 1937+21 (Backer *et al.* 1982). Similarly, a compact source in the supernova remnant CTB80 was pointed out by Strom (1987) as a candidate pulsar; Clifton *et al.* (1987) then showed that this is a pulsar with period 39.5 milliseconds.

Searches for other such short-period pulsars proved to be less rewarding. Manchester *et al.* (1985) using the 210-ft Parkes telescope at 1400 MHz and the Molonglo telescope at 834 MHz found no millisecond pulsars, but discovered instead the interesting young pulsar PSR 1509−58. Eventually, another compact radio source, which was an obvious candidate for a millisecond pulsar, was found in the cluster M28; this source possessed all three characteristics of small angular size, steep spectrum and high polarisation. The source was weak, so that a long integration was needed to reveal the periodicity. This was achieved in 1987, when a Jodrell Bank recording at 1400 MHz was analysed using a Cray XMP super computer at Los Alamos (Lyne *et al.* 1987). The 90-minute recording, which consisted of samples at 300-μs intervals in 32 frequency channels, contained 512 million samples, and the analysis took five hours of computer time. Once the periodicity of 3.05 milliseconds was known, however, routine observations were easily made without the need for a large computer.

4

The distances of the pulsars

4.1 Introduction

The distances of the nearer stars can be obtained from their parallax, which is the apparent annual cyclic movement of position due to the Earth's orbital motion round the Sun. At distances greater than about 1 kiloparsec, stellar distances may be inferred from their apparent brightnesses, since the absolute brightness of a star is generally available from its spectral type.

The distances of some of the closest pulsars may also be obtained from measurements of parallax, with accuracies comparable to the best optical measurements of other stars. The intrinsic brightness of a pulsar, in contrast to that of a visible star of known type, is, however, a very variable quantity, both from time to time and from pulsar to pulsar, so that an observed intensity gives little indication of distance. Fortunately, we have available instead a remarkably direct measurement of distance in the frequency dispersion due to the passage of the radio pulses through the ionised interstellar medium. The magnitude of this dispersion is directly related to the integrated column density of electrons along the line of sight. Hence, the interpretation of measurements of dispersion measures (DM) of pulsars as distances requires a model of the electron distribution in interstellar space; this is itself largely constructed from the measured values of DMs of pulsars, whose distances are known by association with optically identified objects, or by parallax measurements, or by absorption in neutral hydrogen.

In this chapter we assemble the information that is available for the construction of such a model.

4.2 Pulsar distances from parallax

The distances of a small number of pulsars within about 1 kiloparsec of the Sun have been obtained from measurements of their annual

4.3 Pulsar distances from neutral hydrogen absorption

Table 4.1. *Pulsar distances from parallax measurements*

PSR	Parallax (milliarcsec)	Distance (pc)	DM (cm^{-3} pc)	Derived $\langle n_e \rangle$ (cm^{-3})	Reference
0823+26	2.8 ± 0.6	360	19.5	0.054	G 1986
0950+08	7.9 ± 0.8	127	3.0	0.024	G 1986
1929+10	22 ± 8.0	45	3.2	0.071	S 1979
1929+10	<4	>250	3.2	<0.013	B 1982

References: Backer & Sramek 1982; Gwinn *et al.* 1986; Salter *et al.* 1979; Taylor *et al.* 1984.

parallactic motion. This requires a series of positional measurements, spread throughout at least a complete year, with an angular accuracy of about one millisecond of arc. A long baseline radio interferometer is required for such high accuracy.

The first measurements of parallax were made by Salter *et al.* (1979), who used the radio-linked interferometer MERLIN to obtain the distance of PSR 1929+10, and by Backer and Sramek (1981) and Taylor *et al.* (1984) using Very Long Baseline Interferometry (VLBI). The results so far available are shown in Table 4.1. More observations of this type are evidently needed, especially in view of the two discrepant results for PSR 1929+10.

Although the combination of these measured parallaxes and the corresponding dispersion measures provides the most accurate values of the electron density, these are necessarily confined to the regions within 1 kpc of the Sun.

4.3 Pulsar distances from neutral hydrogen absorption

Pulsars at low galactic latitudes may be observed through the spiral arm structure of the Galaxy, and their spectra will then show H I absorption at 21-cm wavelength from structure in front of, but not behind, the pulsar. The velocity structure of this absorption can often be interpreted directly as distance, so that the average $\langle n_e \rangle$ along the line of sight is obtainable from the dispersion measure.

The pulsating nature of the source is a great advantage in this measurement. The spectrum of the 21-cm line as observed directly includes the emission spectrum of hydrogen in the telescope beam, and the superposed absorption spectrum of the pulsar is only a small perturbation. The pulsation means, however, that the perturbation is detectable by comparing the spectrum during the pulses with the spectrum between the pulses. A typical observation integrates this difference over some thousands of pulses. The first observations to be made in this way were of PSR 0329+54 (de Jager *et al.* 1968). Fig. 4.1 shows a comparison of the emission and absorption spectra for four pulsars recorded by Caswell *et al.* (1975) and by Graham *et al.* (1974).

Distances derived from absorption spectra are shown in Table 4.2 (Manchester and Taylor 1981, Manchester *et al.* 1981, Clifton *et al.* 1988). The distances z from the galactic plane are calculated from the galactic latitude b.

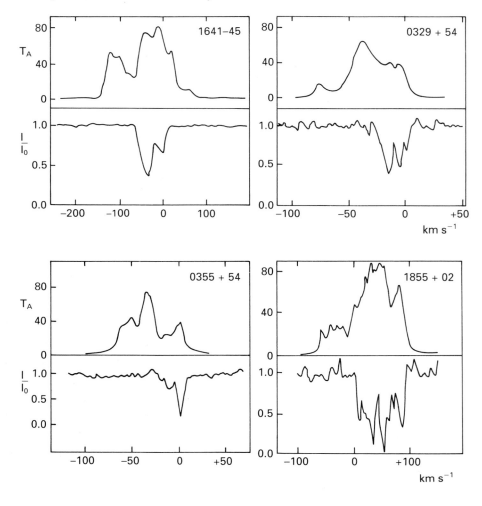

Fig. 4.1. Hydrogen-line absorption spectra in pulsars. The lower trace shows the absorption spectrum, and the upper trace the hydrogen emission spectrum from the same direction. Spectral features that are seen in both absorption and emission correspond to hydrogen gas in front of the pulsar. (For references and pulsar distances see Table 4.1.)

Table 4.2. *Distances from H I absorption*

PSR	DM (cm^{-3})	Distance (kpc)	Reference
0138+59	35	3	Graham *et al.* 1974
0329+54	27	2.3–2.9	Booth & Lyne 1976
0355+54	57	1.0–2.0	Gomez-Gonzales & Guelin 1974
0525+21	51	2	Weisberg *et al.* 1980
0736−40	161	1.5–2.5	Ables & Manchester 1976
0740−28	74	1.5–2.5	Gomez-Gonzales & Guelin 1974
0833−45	69	0.5	Milne 1970
0835−41	148	2.4–5.0	Gordon & Gordon 1975
1054−62	323	12–20	Manchester *et al.* 1981
1154−62	325	3–7	Manchester *et al.* 1981
1240−64	297	4–6	Manchester *et al.* 1981
1323−62	318	5–11	Ables & Manchester 1976
1356−60	295	6–12	Manchester *et al.* 1981
1557−50	270	7–17	Manchester *et al.* 1981
1558−50	170	1.5–2.5	Manchester *et al.* 1981
1641−45	475	4.5–5.3	Ables & Manchester 1976
1849+00	840	8–22	Clifton *et al.* 1988
1859+03	403	6–18	Weisberg *et al.* 1979
1859+07	240	2.5–11	Clifton *et al.* 1988
1900+01	246	3–5	Weisberg *et al.* 1979
1904+06	465	9–17	Clifton *et al.* 1988
1929+10	3	0.04–0.08	Salter *et al.* 1979
2002+31	235	8–13	Weisberg *et al.* 1979
2111+46	142	4–6	Graham *et al.* 1974
2319+60	94	2.8–3.8	Booth & Lyne 1976

4.4 Optically identified pulsars

The Crab Nebula is at a distance of 2 kpc, with an uncertainty of about 20%. The dispersion measure of the pulsar is 56.8 pc cm^{-3}; neglecting the small part of this which may be attributable to ionised gas within the nebula, we immediately obtain a value of 0.028 cm^{-3} for the mean electron density in the line of sight. Similarly, the dispersion measures of pulsars located within globular clusters give values for several different lines of sight, extending our direct measurements to distances greater than 4 kpc. Other optical associations, such as those of the Vela Pulsar PSR 0833−45 and PSR 1509−58 (Manchester *et al.* 1982) with supernova remnants, give less accurate values of $\langle n_e \rangle$ since the distances are less certain (Table 4.3).

Table 4.3. *Pulsar distances from optical identifications*

PSR	Optical object	Distance (pc)	Pulsar DM (pc cm^{-3})	Derived $\langle n_e \rangle$ (cm^{-3})	Reference
0531+21	Crab Neb (SNR)	2000	57	0.028	
0046−71A	47 Tuc (GC)	4600	65	0.014	A 1988
0046−71B	47 Tuc (GC)	4600	65	0.014	A 1988
1620−26	M 4 (GC)	2000	63	0.031	L 1988
1824−21	M 28 (GC)	5800	120	0.021	L 1987
2127+11	M 15 (GC)	9700	58	0.006	W 1989

References: Ables *et al.* 1988; Lyne *et al.* 1987, 1988; Wolszcan *et al.* 1989.

4.5 Interstellar hydrogen

Interstellar matter, which comprises a large proportion of the mass of the Galaxy, is most obvious in the form of dust clouds, which obscure the distant regions of the Milky Way. The more diffuse gas, which pervades the whole Galaxy, is observable optically through the absorption lines of ions such as Ca II, Na I, etc., which appear in stellar spectra. The most abundant element in the interstellar medium is hydrogen. Neutral hydrogen is observed through its radio spectral line at 21 cm wavelength, giving a complete picture of its distribution throughout the Galaxy. Some of this gas is ionised by energetic radiation from stars or cosmic rays. This ionised hydrogen is observable through its continuum radio free-free emission, or bremsstrahlung. At high galactic latitudes most of the observed radio emission is synchrotron radiation from very high-energy electrons; the free-free emission is observed in the hydrogen concentration close to the galactic plane, and especially in the highly ionised H II regions surrounding the bright O and B stars found in the plane. The electron content of this ionised gas, which is distributed throughout the Galaxy, is responsible for the frequency dispersion in the arrival times of radio pulses from the pulsars.

The large-scale distribution of electron density may be considered in two components, which we refer to as the disc and the halo. The shape of the disc is approximately that of the neutral hydrogen disc, although we use a model in which the spiral arm structure is disregarded. Kerr (1969) showed that the broad distribution of neutral hydrogen in the galactic plane is a thin disc with a density n_H, which varies exponentially with distance z from the plane as

$$n_H = n_{H_0} exp(-|z|/z_0).$$

The parameters n_{H_0} (central density) and z_0 (scale height) vary with galactic radial distance R. According to Kerr, the scale height is reasonably constant at 150 pc for $R = 4$–10 kpc, decreasing to 85 pc at the 4-kpc spiral arm, and increasing to 500 pc in the outer arms in the vicinity of $R = 15$ kpc. The central density is about 0.7 cm^{-3} for $R = 7$–11 kpc, decreasing to 0.3 cm^{-3} at $R = 4$ kpc and 0.1 cm^{-3} at $R = 15$ kpc. The detailed distribution of the ionised component will, of course, depend on the distribution of the ionising stars within the hydrogen disc.

Apart from the measurement of pulsar dispersion measures themselves, the ionised component of the hydrogen in this disc is difficult to observe. At radio wavelengths there may be an appreciable optical depth due to free-free collisions; there may consequently be thermal emission, or there may be absorption of radio emission from other more distant sources as observed through the gas. These sources may be discrete extragalactic objects such as quasars, or they may be the diffuse background of the synchrotron radiation from cosmic-ray electrons in the Galaxy. Radio astronomy also allows observation of the ionised gas through the recombination lines, which represent quantum transitions between high-order excited states of hydrogen atoms. These three types of observation, of continuum emission, continuum absorption and line emission, all involve the square of the electron density, and it is important to distinguish between the required average quantity $\langle n_e \rangle$ and the average $\langle n_e^2 \rangle^{1/2}$ measured by these observations. The ratio between these two depends on the clumpiness of the electron distribution; it will also appear that the analysis of the observations depends critically on the temperature. For example, observations of the low-frequency absorption of the extragalactic background at high galactic latitudes suggest that the ionised gas has an optical depth $\tau = 1.5$ at 1 MHz, assuming that the absorption is reasonably uniform over a large solid angle (Alexander, Brown, Clark and Stone 1970). For a layer of scale height 300 pc in the distribution of n_e (and correspondingly with a scale height 150 pc in the distribution of n_e^2), and at a temperature T_3 in units of 1000 K, this optical depth corresponds to

$$\langle n_e^2 T_3^{-1.5} \rangle = 0.0025.$$

The r.m.s. value $\langle n_e^2 \rangle^{1/2}$ would then be 0.05 cm^{-3} for $T = 1000$ K. We note that the pulsar measurements give a more direct measurement, which can be used with the earlier results to place a limit on the actual temperature.

4.6 The H II regions

Within the thin hydrogen disc are the discrete ionised clouds, known as H II regions, surrounding the very hot O and B type stars, which are responsible for their ionisation. A classical analysis by Strömgren

(1936) showed that the ultraviolet light from a hot star ionises a spherical region whose radius S_0 (pc) depends on the density N (cm^{-3}) of the interstellar hydrogen gas, and the radius R (in solar radii) and temperature T (K) of the star, according to

$$\log_{10}(S_0 N^{2/3}) = -0.44 - 4.51 \frac{5040}{T} + \tfrac{1}{2}\log_{10} T + \tfrac{2}{3}\log_{10} R.$$

Within this sphere the ionisation is complete. The density N is somewhat uncertain, and may vary considerably from one H II region to another. It is generally taken to be about 10 cm^{-3}

An individual H II region near the Sun may contribute significantly to the dispersion measure of a pulsar whose line of sight passes through the region. Prentice and ter Haar (1969) and Grewing and Warmsley (1971) have listed these individual regions in the neighbourhood of the Sun and estimated their effect on apparent distances of pulsars. They also estimate the combined effect of the more numerous smaller H II regions; this may be regarded as a minor contribution to the thin disc.

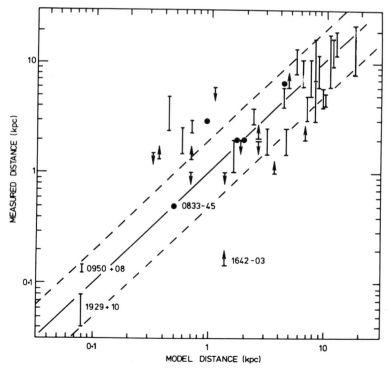

Fig. 4.2. Pulsar distances from the model compared with direct determinations (Lyne, Manchester & Taylor 1985).

4.7 Model electron distribution

Bearing in mind the difficulty of obtaining $\langle n_e \rangle$ from the optical measurements and from the radio emission measures, which provide only $\langle n_e^2 T^{-1.5} \rangle$, we must rely mainly on the pulsar measurements themselves in constructing a model of the electron distribution. The presently accepted model is that of Lyne, Manchester and Taylor (1985). This 'LMT' model aggregates the whole of the thin disc, including all but one of the H II regions, into a single disc component with a scale height of 70 parsecs and a density on the plane (i.e. at $z=0$), which decreases with distance R from the centre of the Galaxy. This variation with galactic radius is adjusted to account for the observed distribution of pulsars with galactic latitude. The Gum Nebula, a very prominent nearby H II region, is treated separately.

There is also a thick disc, with scale height considerably greater than the z distances of most pulsars. The LMT model can only account for this by a component with constant electron density. The model is defined by the equation:

$$n_e = \left[0.025 + 0.015 \exp\left(-\frac{|z|}{70}\right) \right] \left[\frac{2}{1+R/10} \right] + 0.28 \alpha_{(GN)} ,$$

where n_e is in cm^{-3}, the distance from the galactic plane z is in pc, the galactocentric radius is in kpc, and α_{GN} is unity within and zero outside the physical boundaries of the Gum Nebula. The Sun is assumed to be at $R = 10$ kpc, $z=0$.

Some indication of the validity of this model is given by Fig. 4.2, which shows a comparison between distances obtained from the model and distances obtained from the more direct means described earlier in this chapter.

There is very little indication of the electron density distribution at large distances from the plane. Readhead and Duffett-Smith (1975) used observations of diffraction in extragalactic radio sources to show that the scale height is around 500 pc; this value is consistent with the observation that the largest values of dispersion measure for pulsars at high galactic latitudes are about 20 pc cm^{-3}.

We return to the discussion of pulsar distances in Chapter 8, where we consider the population of pulsars in the Galaxy.

5

Pulse timing

Astrophysics provides many examples of rotating and orbiting bodies whose periods of rotation and revolution can be determined with great accuracy. Within the Solar System the orbital motion of the planets can be timed to a small fraction of a second, while the rotation of the Earth is used as a clock that is reliable to about one part in 10^8 per day. Outside the Earth there is, however, no other clock with a precision approaching that of pulsar rotation.

The arrival times of the radio pulses from the pulsars are easy to study, and a surprising amount can be learned from them. Not only do they provide information on the nature of the pulsed radio source, they can also give an accurate position for the source; and they can give information on the propagation of the pulses through the interstellar medium. All three kinds of information were noted by Hewish and his collaborators in the discovery paper of 1968. They showed that the shortness of the pulses, and their short and precise periodicity, implied that the source was small, and that it might be a rotating neutron star. They showed also that the arrival time was varying because of the Doppler effect of the Earth's motion round the Sun; this annual variation implied that the source lay outside the Solar System. Finally, they showed that the arrival time of a single pulse depended on radio frequency; this dispersion effect was found to be in accord with the effect of a long journey through the ionised gas of interstellar space.

More recently, arrival time measurements have provided precise information on the binary motion of several pulsars, made possible some fundamental tests of general relativity and gravitational radiation, and provided time standards with stability challenging that of the best atomic clocks.

5.1 Pulsar positions and the Earth's orbit

Since the time of Römer, who made observations of the motion of Jupiter's satellites when the Earth was at different positions in its orbit, it

Pulsar positions and the Earth's orbit

has been known that light takes about 8½ minutes to travel from the Sun to the Earth. Pulses from a pulsar lying in the plane of the ecliptic will therefore arrive earlier at the Earth than at the Sun when the Earth is closest to the pulsar, i.e. when it is at the same heliocentric longitude. Six months later the pulses will arrive late by the same amount. Assuming for simplicity that the Earth's orbit is circular and centred on the Sun, the delay t_c is given by

$$t_c = A \cos(\omega t - \beta) \cos \lambda \tag{5.1}$$

where A is the light travel time from Sun to Earth, ω is the angular velocity of the Earth in its orbit, and β, λ are the ecliptic longitude and latitude of the pulsar (Fig. 5.1a).

The observed arrival times of pulses emitted by a pulsar at equal time intervals throughout the year will therefore show a sinusoidal variation as in Fig. 5.1b, where the phase gives the heliocentric longitude and the

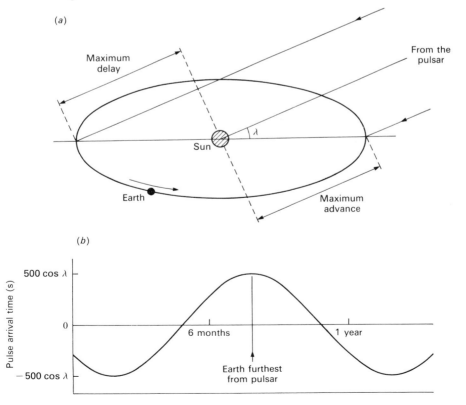

Fig. 5.1 (*a*) The annual variation in pulse arrival time due to the Earth's orbital motion round the Sun. (*b*) The amplitude of the variation is 500 cos λ s, where λ is the ecliptic latitude of the pulsar. The phase of the sinusoid is used to determine longitude.

amplitude gives the heliocentric latitude of the pulsar. The angular accuracy of the positions so determined is greatest near the pole of the ecliptic, since the ecliptic latitude λ is poorly determined near $\lambda=0$. An error in coordinates $\delta\lambda$, $\delta\beta$ gives rise to periodic timing errors

$$\delta t_c = -A\delta\lambda \cos(\omega t - \beta)\sin\lambda + A\, \delta\beta \sin(\omega t - \beta)\cos\lambda. \tag{5.2}$$

Positions obtained from the variation of pulse arrival times through a year are remarkably accurate. Typically, an observation of a pulsar with a period of about 1 s can give a point on the timing curve with an accuracy of about 0.2 ms. At least four such observations through one year are needed to find the source position, but the result is a position accurate to the order of 0.1 arc second. This is sufficiently accurate for any attempts at identification with unusual visible or X-ray objects, and positions have therefore been obtained from timing observations for most of the known pulsars. The application of the method requires consideration of the following details (Hunt 1971):

(i) The pulsar period lengthens as the rotation rate of the pulsar slows down during the observation, adding a parabola to the curve of Fig. 5.1. (Determination of the size of this parabola may, of course, be a prime objective of the observations, since it determines $\dot{P}$, which gives a measure of pulsar age.)

(ii) Rotation of the Earth introduces a variable time delay of up to the transit time over one Earth radius (21 ms).

(iii) The Earth's orbit is elliptical, not circular.

(iv) The Sun moves in relation to the centre of inertia of the Solar System, known as the barycentre, which is a convenient inertial frame of reference. This motion depends on the orbital motion of the planets, mainly the massive planet Jupiter; it is sufficiently large to place the barycentre just outside the surface of the Sun.

(v) The gravitational potential at the Earth differs from the potential at a large distance from the Sun; furthermore it varies annually through the ellipticity of the Earth's orbit. General Relativity therefore predicts a small annual variation of Earth-bound clock rate, as compared with a reference clock in a circular orbit.

(vi) The second-order Doppler effect, predicted by Special Relativity, varies as the square of the Earth's velocity: the variation of this effect due to the ellipticity of the orbit is in practice indistinguishable from the general Relativistic effect in (v) above.

(vii) The effective frequency of the radio receiver as observed in an inertial frame of reference varies through the year due to the Doppler

5.2 The barycentric correction

effect of the Earth's motion. Since the arrival time depends on frequency, due to dispersion in the interstellar medium, a correction may be needed for timing observations of pulsars with high dispersion measure.

5.2 The barycentric correction

In practice the expected arrival times are computed using an assumed position β, λ and an ephemeris giving the vector distance $\mathbf{r}_{ob}$ from the observer to the Solar System barycentre. This vector distance is the sum of the three vectors, $\mathbf{r}_{oc}$ from the observer to the centre of the Earth, $\mathbf{r}_{cs}$ from the centre of the Earth to the centre of the Sun, and $\mathbf{r}_{sb}$ from the centre of the Sun to the barycentre:

$$\mathbf{r}_{ob} = \mathbf{r}_{oc} + \mathbf{r}_{cs} + \mathbf{r}_{sb}. \tag{5.3}$$

Then the time t_c to be added to the observed time to give a barycentric arrival time is

$$t_c = -\frac{1}{c} \mathbf{r}_{ob} \cdot \mathbf{s} \tag{5.4}$$

where $\mathbf{s}$ is the position vector of the source at β, λ.

The three components of $\mathbf{r}_{ob}$ are obtained separately. The vector $\mathbf{r}_{cs}$ is by far the largest and is available in terms of the astronomical unit; in practice it may be necessary to combine two ephemerides that give respectively the motion of the barycentre of the Earth–Moon system and the motion of the Earth within that system. The astronomical unit itself has been well determined as a light travel time from planetary radar observations. The barycentric correction $\mathbf{r}_{sb}$ is obtained from the vector positions, $\mathbf{r}_i$ and masses m_i of the Sun and planets (measured in units of the solar mass):

$$\mathbf{r}_{sb} = \frac{1}{1 + \Sigma \frac{1}{m_i}} \Sigma \frac{1}{m_i} \mathbf{r}_i. \tag{5.5}$$

The Earth radius correction $\mathbf{r}_{oc}$ may be computed directly as a light travel time, since it depends only on the radius of the Earth at the observatory and the source elevation E. At the mean radius of the Earth the time correction is $21.2 \sin E$ ms.

The two ephemerides in common use for pulsar astrometry are derived from an analysis of planetary radar results. Radar measurements improved the accuracy of the astronomical unit by two orders of magnitude (Ash, Shapiro and Smith 1967), providing the 'MIT' ephemeris (named after the Massachusetts Institute of Technology), which is accurate to about 10 microseconds in planetary distances and 5 microseconds in the Earth–Sun distance. The astronomical unit is quoted as 499.004786 ± 5 light seconds. An independent ephemeris (Standish 1982), known as the 'JPL' ephemeris (named after the Jet Propulsion Laboratory) is also in use.

5.3 The general relativistic correction

The annual variation of an atomic clock on the surface of the Earth as it follows its elliptical orbit round the Sun was analysed by Clemence and Szebehely (1967). The difference between the time S shown by a clock on the Earth, and coordinate time t shown by an identical clock at an infinite distance from the Sun is given by

$$\frac{dt}{dS} = 1 + \left(\frac{1}{r} - \frac{1}{4a}\right) 2GMc^{-2} \tag{5.6}$$

where r is the Earth–Sun distance, a is the semi-major axis of the Earth's orbit, G is the gravitational constant and M is the mass of the Sun. The constant $2GMc^{-2} = 2.95338$ km.

The major part of this difference, amounting to a rate of 1.48×10^{-8}, is incorporated into the definition of atomic time, which refers to a standard clock in orbit at a constant distance a from the Sun. The variable part, due to the variation of r round the Earth's orbit, is a rate amounting to $3.3079 \times 10^{-10} \cos f$, where f is the 'true anomaly' i.e. the angle between the least radius vector of the Earth's orbit (at perihelion) and the instantaneous vector. Since the orbit is elliptical, the true anomaly is slightly different from the 'mean anomaly' l, which increases uniformly with time. The integrated fractional difference between the atomic clock and the standard clock at any time is given by the integral

$$3.3079 \times 10^{-10} \int_0^l \cos f \, dl,$$

which contains a major term $\sin l$ and minor terms depending on the eccentricity e:

$$\int_0^l \cos f \, dl = \sin l + e(\tfrac{1}{2}\sin 2l - l) + \tfrac{3}{8}w^2(\sin 3l - 3 \sin l). \tag{5.7}$$

The term $-el$ represents a constant rate error, and is therefore absorbed into the definition of the standard clock; the result is a relativistic correction Δt_r given to the nearest microsecond by:

$$\Delta t_r = 0.001\,661 \sin l + 0.000\,028 \sin 2l \text{ seconds} \tag{5.8}$$

The clock is right on January 1, when $l=0°$; the maximum error is approximately at $l = \pm 89°$, i.e. on April 1, when the clock is fast, and on October 30 when the clock is slow. A more precise analysis is presented by Blandford and Teukolsky (1976).

A further general relativistic effect concerns the time for a radio pulse to traverse the gravitational potential well of the Solar system. The gravitational potential of the Sun introduces a delay, which reaches a maximum of 250 microseconds when the line of sight is close to the Sun's limb. A full discussion of this and other relativistic effects in pulsar timing is given by Backer and Helling (1986).

5.4 The fundamental reference frame

Astrometric calculation based on the annual variation of pulse arrival times yields pulsar positions in ecliptic coordinates, which are related to the Earth's orbit round the Sun. Conventional radio interferometer techniques yield positions in equatorial coordinates, which are based on the rotation of the Earth. The poles of these two coordinate systems are inclined at the obliquity angle of 23½°, and the intersection of their equatorial planes defines the direction of the vernal equinox. Their exact relationship is believed to be known to about 0.1 arcsecond. A comparison of interferometric and timing positions of pulsars can in principle be used to check this relationship.

This comparison was made for 59 pulsars with good timing positions and with interferometric positions measured on the VLA (the Very Large Array in New Mexico), with an accuracy of about 0.2 arc second (Fomalont *et al.* 1984). Most positions agreed within the combined measurement errors, and the mean offsets between the two coordinate systems were shown to be less than 0.2 seconds. This agreement conceals some appreciable systematic differences concentrated in particular areas of the sky, and there is now considerable interest in improving the accuracy of the comparison to the point where pulsar astrometry can contribute significantly to the fundamental coordinate systems.

5.5 Period changes

All known pulsars, except the X-ray pulsars, have the basic characteristics of an intrinsically precise period, modulated only by a slow monotonic increase in period due to a gradual loss of rotational energy. The change of period of the Crab Pulsar can be detected within a few hours, and the change for the Vela Pulsar within days, but generally the rate of change is so small that it can only be determined from observations over a period of one year. Furthermore, accurate positions of pulsars are usually only available from the timing observations themselves, so that an apparent change of period over a short observing time may be due only to an error in the assumed pulsar position. It may, however, be an indication that the pulsar is in a binary system.

The rate of change of period, $\dot{P}$, is an important characteristic of a pulsar. It is conveniently found by measuring the period P at two epochs close to one year apart; the result is then independent of small errors in position. More generally, a series of observed pulse arrival times distributed throughout the year may be used to obtain values of P, $\dot{P}$ and position by a least-squares fitting procedure. This is conveniently done in terms of the pulsation frequency $\nu = P^{-1}$; starting at an arbitrary time t_0 the expected pulse number N at an observed arrival time t is given by

$$N = \nu_0(t-t_0) + \tfrac{1}{2}\dot{\nu}(t-t_0)^2 + \tfrac{1}{6}\ddot{\nu}(t-t_0)^3. \tag{5.9}$$

For assumed values of ν_0, $\dot{\nu}$ and $\ddot{\nu}$, the expected values of N are computed for the observed arrival times t, corrected to the barycentre using an assumed polar position. Generally N is not then integral; the fitting process is the adjustment of the parameters to minimise the fractional part of N over the whole run of observations.

The outstanding result from extended timing measurements on many pulsars is that the arrival times of the pulses are astonishingly regular. Discontinuities in arrival time, period P or its differential $\dot{P}$ have been observed in only a few pulsars, notably the Vela Pulsar and the Crab Pulsar; others show a low level of timing irregularity referred to as timing noise (Chapter 6).

5.6 Pulsar ages and the braking index

According to classical electrodynamics, a magnetic dipole with moment $M_\perp$, rotating at angular velocity Ω about an axis perpendicular to the dipole, radiates a wave at angular frequency Ω with a total power $\tfrac{2}{3}M_\perp^2\Omega^4 c^{-3}$. The energy supply is the angular kinetic energy of the rotating body. Let the moment of inertia be I. Then

$$d(\tfrac{1}{2}I\Omega^2)/dt = I\Omega\dot{\Omega} = \tfrac{2}{3}M_\perp^2\Omega^4 c^{-3}. \tag{5.10}$$

The energy flow from a pulsar may be a combination of this dipole radiation and an outflow of particles. Even if the particles carry a large share of the energy, the total energy flow is approximately given by equation (5.10), since the magnetic field dominates the physics of the outer magnetosphere. The combined outflow may be regarded as a flow at velocity c through a area $4\pi R_c^2$, where R_c is the radius of the velocity-of-light cylinder. Provided that the total energy density at the velocity-of-light cylinder is approximately $B^2/8\pi$, the flow is given by equation (5.10); we

5.6 Pulsar ages and the braking index

will see, however, that the observed slowdown law is significantly different, so that this argument must be somewhat oversimplified.

The dipole moment M can be derived from measured values of P and $\dot{P}$. A conventional value of the polar field at the surface, B_0, is often quoted: this assumes an orthogonal rotator with a radius 10 km and moment of inertia 10^{45} g cm^2, giving

$$B_0 = 3.3 \times 10^{19} (P\dot{P})^{1/2} \text{ gauss.} \tag{5.11}$$

If we assume that a pulsar is formed with a high angular velocity Ω_i, subsequently evolving according to a simple power law

$$\dot{\Omega} = -k\Omega^n$$

where k is a constant and n is referred to as the 'braking index', we obtain by integration the relation between Ω, $\dot{\Omega}$ and t:

$$t = -\frac{\Omega}{(n-1)\dot{\Omega}} \left[1 - \frac{\Omega^{n-1}}{\Omega_i^{n-1}}\right] \tag{5.12}$$

Provided $n \neq 1$, and $\Omega_i \gg \Omega$, we can approximate to a characteristic age τ given by

$$\tau = -\frac{1}{n-1} \frac{\Omega}{\dot{\Omega}} = \frac{1}{n-1} \frac{P}{\dot{P}}. \tag{5.13}$$

For magnetic dipole braking, $n=3$, giving a characteristic age $\tau = \frac{1}{2}P/\dot{P}$; this is the accepted definition of characteristic age, even though the value of n may differ from 3.

A direct measurement of n is obtainable only if the second differential $\ddot{P}$ can be found. In terms of angular velocity Ω or frequency $\nu = \Omega/2\pi$:

$$n = \frac{\Omega\ddot{\Omega}}{\dot{\Omega}^2} = \frac{\nu\ddot{\nu}}{\dot{\nu}^2} = 2 - \frac{P\ddot{P}}{\dot{P}^2}. \tag{5.14}$$

Significant values have been found for the Crab Pulsar ($n = 2.515 \pm 0.005$, Lyne, Pritchard and Smith 1988) for PSR 1509−58 ($n = 2.8 \pm 0.2$, Manchester, Durdin and Newton 1985) and for PSR 0540−69 ($n = 2.01 \pm 0.02$, Manchester and Peterson 1989). The deviations from the expected value $n=3$ for purely magnetic dipole radiation indicate that part of the torque on the pulsar is due to outflow of particles.

Sufficient timing data now exist for the Crab Pulsar to allow a measurement of the third derivative $\dddot{\Omega}$ to an accuracy of about 10% (Lyne, Pritchard and Smith 1988). The value agrees with the theoretical relation

$$\dddot{\Omega} = \frac{n(2n-1)\dot{\Omega}^3}{\Omega^2}. \tag{5.15}$$

54 Pulse timing

If all pulsars were identical at birth, having the same magnetic fields and moment of inertia, and their rotation slowed down according to the power law with braking index 3, there would be a unique relation between P and $\dot{P}$ such that a logarithmic plot would show a straight line with slope -1. In practice the plot shows a wide scatter (Fig. 5.2), with no obvious correlation between P and $\dot{P}$. Obviously each point in this plot is pursuing a track sloping downwards in the direction of increasing P and decreasing $\dot{P}$, but the scatter of points indicates that the individual tracks are very different. Later we discuss the decay of the magnetic dipole moment, which affects the later stages of evolution through this so-called P–$\dot{P}$ diagram.

5.7 Pulsars as standard clocks

The discovery of the millisecond pulsars (Chapter 10), all of which have very low values of $\dot{P}$ and very low timing noise, opens up a new possibility for a standard of time. The present definition of Universal Time

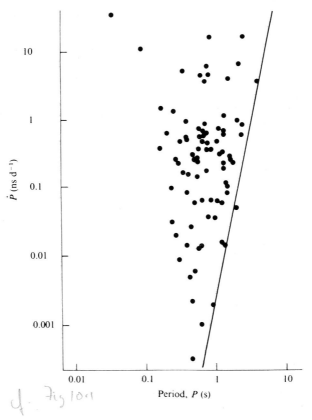

Fig. 5.2. The P–$\dot{P}$ diagram: logarithmic plot of period against its time derivative. This early plot was made before the discovery of the millisecond pulsars.

5.7 Pulsars as standard clocks

(UTC) is in terms of an ideal caesium clock, but in practice it is realised as the average of a selected set of caesium clocks. A standard of time based on pulsars would have no link to a reproducible physical phenomenon such as the oscillation of a caesium atom, but it may nevertheless provide a smoother running clock than the present UT.

The smooth running of a clock is characterised by the Allan variance of its errors, which is a measure of its fractional stability. Fig. 5.3 shows that this typically decreases over a period of time during which the behaviour of the clock is predictable; for a standard caesium clock this is about one month. The pulsar PSR 1937+21 has been observed over a period of several years by Davis et al. (1985); although, on a short time scale, this pulsar clock cannot be read as accurately as others shown in Fig. 5.3, the fractional accuracy continues to improve with time and is likely to overtake all others. The short-term irregularities of the pulsar clock appear to be random, with an r.m.s. amplitude of about 1 µs, as shown in Fig. 5.4.

The possibility of creating a useful pulsar time scale can only be checked through an extensive series of observations of several of the smoothest running pulsars that might be combined to give a 'mean pulsar clock'. If it can be shown that the differences between pairs of pulsar clocks (after making due allowance for the various values of period and period derivative) are smaller than the differences between UT and the mean pulsar

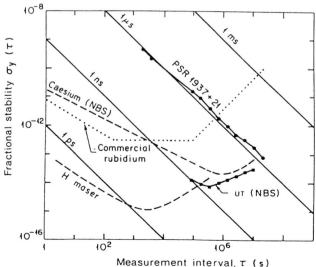

Fig. 5.3. Measured frequency stability $\sigma(\tau)$ for PSR 1937+21, over intervals τ from 30 min to 256 days. The performances of caesium, rubidium, and H-maser clocks are shown for comparison (Davis et al. 1985).

clock, then a new timescale could be defined based on pulsars alone. The practical application of such a time scale would initially be to improve the smooth running of UT itself; a re-definition of time intervals that abandons caesium clocks altogether seems unlikely.

5.8 Proper motion

The proper motions of several pulsars are large enough to be measured by observing pulse arrival times over a period of several years. This was first achieved for PSR 1133+16 by Manchester *et al.* (1974), using timing observations over a four-year period. Fig. 5.5 shows the progressively increasing error in pulse arrival times due to the proper motion (Helfand *et al.* 1977).

This method of measuring proper motion has proved to be difficult and unreliable because of the random timing noise observed in most pulsars (Chapter 6). Direct interferometric measurements such as those by Lyne, Anderson and Salter (1982) have yielded most of the known values of proper motion.

A large transverse component of velocity gives rise to an appreciable increase in period P even if the pulsar is not slowing down. Let the transverse velocity be V, and the distance r. Then the apparent rate of change is

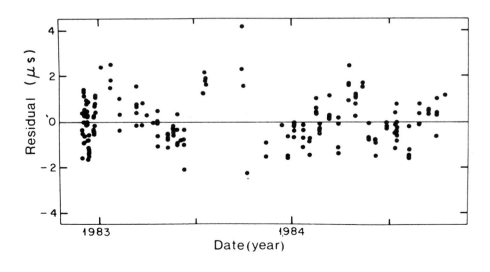

Fig. 5.4. Timing residuals for PSR 1927+21 after fitting the observations for best-fit values of period, period derivative, position, and proper motion. Each point represents a 30-minute observation (Davis *et al.* 1985).

5.9 Binary pulsar systems

$$\dot{P} = \frac{V^2}{rc} P.$$

As a measure of the possible importance of this effect, we can estimate the characteristic time $\frac{1}{2}P/\dot{P}$ and compare it with observed values. For example, a pulsar at the rather small distance of 50 pc, with a velocity of 200 km s^{-1} would have a characteristic time of 10^8 years. For most pulsars the measured values of $\frac{1}{2}P/\dot{P}$ are of order 10^6 to 10^7 years, so that the effect is unimportant. However, for PSR 1133+16 it amounts to 5% of the measured value of $\frac{1}{2}P/\dot{P}$.

5.9 Binary pulsar systems

The pulse arrival times for most pulsars are notably free from the modulation that is typically found in the pulsating X-ray sources, and which is caused by the orbital motion in a binary system. Only about two per cent of the known pulsars are obviously in binary systems; we examine these in detail in Chapter 10. Here we examine briefly the possibility that undetected orbiting companions may exist in some of the apparently solitary pulsars.

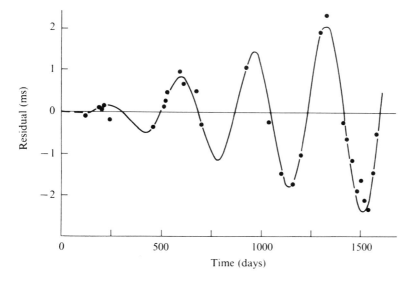

Fig. 5.5. Proper motion of PSR 1133+16. The growing sinusoidal pattern of errors in pulse arrival time is due to an angular motion of about ½ arcsecond per year (Manchester *et al.* 1974).

A heavy planet at a large distance from a pulsar should be detectable because of the periodic displacement of the pulsar from the barycentre of the system. Jupiter, for example, orbiting with a period of 11.9 years, causes the Sun to be displaced from the barycentre of the solar system by 2.5 light seconds. Lighter planets at larger distances would evidently go undetected at least for several years.

Binary star systems are easily detected if the periods are shorter than a few years. Longer period binaries are certainly not common; if they were, at least some pulsars would be showing a negative value of $\dot{P}$. Limits on possible numbers of long term binaries are hard to establish but, as a general rule, the orbital velocities cannot be such as to give apparent characteristic ages $\frac{1}{2}P/\dot{P}$, less than 10^6 year. Undetected binary systems with periods of several decades may indeed exist. The range of possibilities is, of course, diminishing as timing observations extend over longer intervals.

5.10 The binary pulsar PSR 1913+16

Well over a hundred pulsars had been discovered before one was found to be a member of a binary system. Several are now known (Chapter 10), but the first is so remarkable that it is often referred to as The Binary Pulsar. This pulsar, PSR 1913+16, was discovered by Hulse and Taylor (1974) during a systematic search using the Arecibo radio telescope (Chapter 2). The orbital period of 7¾ hours is remarkably short – so short, in fact, that the rapid changes of period due to changing Doppler shift made its detection and confirmation particularly difficult. The period changes through a range of more than one part in a thousand during the orbit, so that successive observations showed changes in Doppler shift well outside the accuracy of measurement even in the initial observations. Fig. 5.6 shows the velocity curve obtained from early observations. No eclipsing occurs.

This velocity curve can immediately be interpreted in terms of an elliptical orbit, with the following main characteristics:

Period (P_b) = 27907 s
Projected semi-major axis ($a \sin i$) = 7.0×10^5 km
Orbital eccentricity (e) = 0.617.

The extraordinary character of this orbit becomes apparent when it is realised that the measured velocity reaches 10^{-3} of the velocity of light.

5.10 The binary pulsar PSR 1913+16

The actual velocity, and the radial distances from the centre of mass of the binary system, depend on the inclination i of the plane of the orbit to the plane of the sky and will be even greater. Anticipating later discussion we set $\sin i = 0.8$; it follows that the orbital velocity varies from $1.33 \times 10^{-3} c$ to $0.316 \times 10^{-3} c$, and the radial distance from 3.3×10^5 to 14×10^5 km. The whole orbit would fit within the diameter of the Sun.

Such an orbit is expected to show major special relativistic effects, and appreciable general relativistic effects. Fortunately the pulsar has a very short and stable period and a slow rate of slowdown ($P = 59$ ms, $\dot{P} = 8.6 \times 10^{-18}$), so that an ideal clock is following the orbit and can be observed with an accuracy of about 50 microseconds by averaging pulse arrival times over a period of 5 minutes. The results of observations over a period of seven years are most exciting, demonstrating several relativistic effects (Taylor and Weisberg 1982).

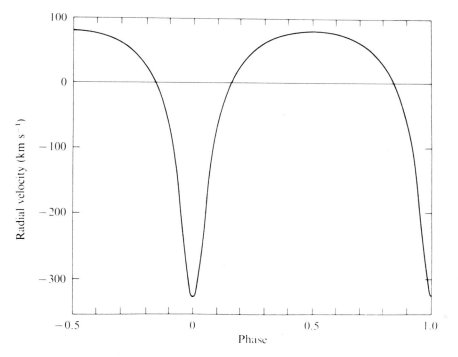

Fig. 5.6. Radial velocity curve for the binary pulsar PSR 1913+16. The velocity is found from the modulation of the pulse period due to the Doppler effect. The curve is markedly non-sinusoidal, indicating the large eccentricity of the orbit (Hulse & Taylor 1974).

5.11 Relativistic effects in the binary orbit

The arrival times of pulses from PSR 1913+16 require correction for the position of the observatory relative to the barycentre of the Solar System, as described in equation (5.1). Similar corrections apply to the orbit of the pulsar in the binary system. The propagation times across the orbit are of order v/c times the orbital period. Two extra terms of order v^2/c^2 are important: one is due to the varying gravitational potential, as in equation (5.6), and the other is the transverse Doppler shift, which affects clock rate due to a velocity transverse to the line of sight. Finally, there is a propagation delay across the orbit; this is familiar as a delay in propagation time across the Solar System, where space probe signals have been observed to be delayed by 250 microseconds when the line of sight passes close to the Sun (see Section 5.3). This delay provides one of the classic tests of the Theory of General Relativity. In the binary pulsar, the effect is of order $(v/c)^3 P_b$, i.e. of order 50 microseconds. All these effects on propagation time and clock rate have been clearly demonstrated. We now turn to some further effects, which concern the geometry of the orbit.

The earliest classical test of General Relativity was through accurate measurement of the precession of the orbit of the planet Mercury. The size of this effect depends on the strength of the gravitational field in the orbit. For Mercury the component of the rate of advance of periastron (i.e. the movement of major axis of the elliptical orbit) due to General Relativity is 43 arcseconds per century. The crucial factor is the ratio between rest mass energy and gravitational potential; for PSR 1913+16 this factor ($GM/c^2 r$) reaches 10^{-6} and the precession is 4.2 degrees per year.

This measured value of the rate of advance of periastron ($\dot{\omega}_{GR}$) is important, since it depends on the total mass of the system. If m_p and m_c are the masses of the pulsar and its companion, then

$$\dot{\omega}_{GR} = 2.11 \left(\frac{m_p + m_c}{M_\odot}\right)^{2/3} \text{ degree yr}^{-1}.$$

It follows from the measured value that $m_p + m_c = 2.8\ M_\odot$, provided that there is no contribution from tidal or rotation effects in the companion star. Both components are evidently neutron stars, so that such effects are negligible.

The various effects on the pulse arrival time vary round the orbit in different ways, so that a long series of precise timings can be used to find a consistent set of orbital parameters, including the orbital inclination. The fit of 1000 observations spanning four years even allows a separate determination of m_p and m_c, since the different relativistic effects are pro-

5.11 Relativistic effects in the binary orbit

portional to different combinations of the masses. In this way, Weisberg and Taylor (1984) find that the best fit gives $m_p = m_c = 1.4$ M$_\odot$ and sin i = 0.76 ± 0.14.

Two further effects predicted by General Relativity are now within reach of the observations. The orbital energy of the binary is expected to diminish through gravitational radiation, giving a decrease of orbital period as the stars spiral in. This has produced a cumulative effect of 2 seconds in the phase of the orbit over a period of 6 years (Fig. 5.7), giving a rate of change of period

$$\dot{P}_b = (-2.40 \pm 0.09) \times 10^{-12}.$$

This is consistent with the expected rate of energy loss through gravitational radiation and constitutes the first demonstration of the existence of such radiation. The second effect, which is not so clearly demonstrated as yet, is a coupling of the angular momentum of the rotating pulsar with the angular momentum of the orbit. This should cause the rotation axis of the pulsar to precess, with the interesting consequence that the observer's line of sight may make a different cut across the beam of radiation. Taylor *et al.* show that there have indeed been small changes in the integrated pulse shape, which may be attributed to precession in the pulsar. If this is

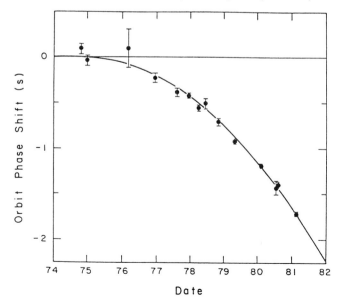

Fig. 5.7. The effect of gravitational radiation on the orbit of the binary pulsar PSR 1913+16. The deviation from constant orbital period is apparent as a cumulative change of orbital phase (Weisberg & Taylor 1984).

Table 5.1. *Rotation and orbital parameters of PSR 1913+16*

Right Ascension (1950.0)	$\alpha =$	$19^h53^m12^s.474 \pm 0^s.004$
Declination (1950.0)	$\delta =$	$16°01'08''.06 \pm 0''.02$
Period	$P =$	$0.05902\,000269 \pm 2$ s
Period derivative	$\dot{P} =$	$(8.64 \pm 0.02) \times 10^{-18}$
Projected semi-major axis $a_1 \sin i =$		$2.3418 \pm .0001$ light seconds
Orbital eccentricity	$e =$	$0.617\,127 \pm 0.000\,003$
Binary orbit period	$P_b =$	$27\,906.98163 \pm 0.000\,02$ s
Longitude of periastron	$\omega_o =$	$178.864 \pm 0.001°$
Rate of advance of periastron	$\dot{\omega} =$	4.2263 ± 0.0003 deg yr^{-1}
Derivative of orbit period	$\dot{P}_b =$	$(-2.40 \pm 0.09) \times 10^{-12}$

These quantities are taken from Taylor & Weisberg (1982) and from Weisberg & Taylor (1984).

confirmed, there is the possibility not only of a full delineation of the shape of the beam in two dimensions, but also that precession will take the beam direction so far from the line of sight that the pulsar becomes unobservable.

The accuracy of measurement of the rotation period and orbital parameters is demonstrated in Table 5.1. The indications are that the binary system is acting very close to the ideal of two point masses, and that it is therefore providing the classic text-book test of the effects of Special and General Relativity on the orbit of a binary system.

6

Timing irregularities

6.1 Timing noise and glitches

On a time scale of some days, all pulsars show a remarkable uniformity of rotation rate. This is not, however, a surprising observation since uniform rotation is exactly what is expected of a spinning body with a large moment of inertia, isolated in space. There is, of course, the slowdown torque of the magnetic dipole radiation, causing the steady increase in period; no other external torque can be expected unless there are changes in the magnetosphere producing irregularities in the dipole field or in the material outflow. Nevertheless, some very interesting irregularities in pulsar rotation have been observed, which are related to changes in the interior of the neutron star.

These internal changes in pulsars can be related to the steady slowdown. Perhaps the most obvious theoretical connection is through the changes in equilibrium ellipticity of the crust that must occur during the slowdown as the centrifugal force reduces. The solid crust of the pulsar must adjust to the changing ellipticity in a series of steps. The resulting changes in moment of inertia might be large enough to account for the steps in rotation speed. Another possible internal change, which seems to account for most of the observations, concerns the independent motion of the crust and the fluid interior. In either case the observations provide a remarkable insight into the pulsar interior, and we will therefore describe them in some detail.

Two categories of timing irregularities exist: a general 'noisy' and fairly continuous erratic behaviour, and a more spectacular step change in rotation speed, popularly known as a 'glitch'. Glitches are rare; they have been observed in only 10 pulsars. They occur more frequently in the younger pulsars; half of those observed so far have been in the Crab and Vela Pulsars. No glitches have been observed in the millisecond pulsars, which are evolving more slowly than the majority of pulsars. Timing noise

64   *Timing irregularities*

is more widespread, although it again is more prominent in the younger pulsars. We describe the glitches first.

6.2   The Vela Pulsar PSR 0833−45

This pulsar has a period of 89 ms; the only other 'young' pulsar with a shorter period is the Crab Pulsar. The rate of change of period is easily measurable and amounts to 10.7 ns/day; it corresponds to the comparatively short lifetime (about 10^4 years) known to be associated with the Vela supernova remnant.

Between observations made on 24 February and on 3 March 1969, the period of this pulsar decreased unexpectedly by about 20 ns, subsequently

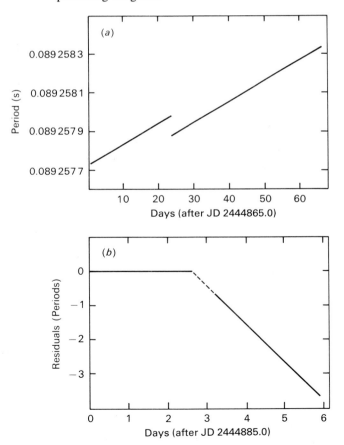

Fig. 6.1. The glitch in the Vela Pulsar, October 1981 (McCulloch *et al.* 1983). (*a*) The mean period of the pulsar from daily observations. (*b*) Timing residuals (in units of one period) using a period obtained from the three days immediately preceding the glitch.

recovering over several weeks to a period close to, but not identical with, that expected from earlier observations. The step change in rotational frequency, amounting initially to two parts in 10^6, probably occurred very rapidly: subsequent observations of other such steps suggest that they occur in less than one day. The partial recovery is much slower, in the form of an exponential with time constant of about one year.

Five such steps occurred between the discovery of the Vela Pulsar in 1968 and October 1981, when the observations shown in Fig. 6.1 were made. These are the daily measurements of period from Hobart, supplemented by weekly measurements from the NASA Deep Space Tracking Station at Tidbinbilla in Australia (McCulloch *et al.* 1983). Fortunately, the daily monitoring had already started when the step occurred; it must have been very gratifying for the observers to see it happen only two days after they started, since the interval between steps is randomly spread between two and three years.

The step of October 1981 corresponded to about one extra rotation per day. Timing observations can be made to an accuracy of about 10^{-4} of the period, so that the progress of the recovery from the step can be very precisely monitored. Analysis was first directed at fitting the pre-step data to the best value of rotational frequency v and its first and second derivatives. Then an exponential recovery was fitted to the post-step data, and the timing residuals were plotted, as in Fig. 6.2. Here the main recovery has already been allowed for: the best fit gave a time constant of 233 ± 1 days, consistent with the early observations, and a recovery of 17.6% of

Fig. 6.2. The rapid exponential recovery from the glitch in the Vela Pulsar, October 1981 (McCulloch *et al.* 1982).

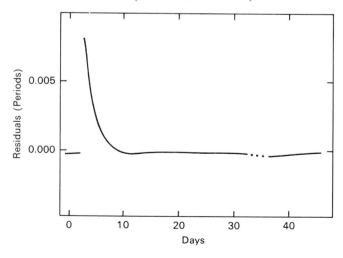

Table 6.1. *Glitches: Crab and Vela Pulsars*

PSR	Date & reference	MJD	Initial jump $\frac{\Delta\nu}{\nu}$	$\frac{\Delta\dot{\nu}}{\dot{\nu}}$	Recovery τ_{days}	Q	$\left(\frac{\Delta\nu}{\nu}\right)_f$	$\left(\frac{\Delta\dot{\nu}}{\dot{\nu}}\right)_f$
0531+21 (Crab)	1969 Sept B1972	40494	$\sim 1\times 10^{-8}$		~ 4			
	1975 Feb DP1983	42448	$\sim 4\times 10^{-8}$	$\sim 10^{-3}$	10.4	~ 0.7	1.2×10^{-8}	2.1×10^{-4}
	1986 Aug LP1987	46664	0.92×10^{-8}	2.5×10^{-3}	5.5	1	0	0
0833−45 (Vela)	1969 Feb RM1969, RD1969	40280	2.3×10^{-6}	8×10^{-3}	400	~ 0.15		
	1972 Aug RD1971	41192	2.0×10^{-6}	$\sim 10^{-2}$				
	1975 Sept	42683	2.0×10^{-6}	7.5×10^{-3}	500	0.22		
	1978 July M1983	43692	3.1×10^{-6}		400			
	1981 Oct McC1983	44888	1.1×10^{-6}	7.2×10^{-3} (5×10^{-2})	233 (and 1.6)	0.18		$<10^{-4}$
	1982 Aug C1988	45192	2.0×10^{-6}	1.0×10^{-3} (5×10^{-2})	60 (and 3.2)	0.04		
	1985 July McC1987	46258	1.6×10^{-6}	6.2×10^{-3} (1×10^{-2})	397 (and 6.1)	0.17		
	1988 Dec F1988	4752;	1.8×10^{-6}	8×10^{-2}	—	—		

References: Boynton *et al.* (1972); Demianski & Proszynski (1983); Lyne & Pritchard (1987); Radhakrishnan & Manchester (1969); Reichley & Downs (1969); Reichley & Downs (1971); Manchester *et al.* (1983); McCulloch *et al.* (1983); McCulloch *et al.* (1987); Flanagan (1988); Cordes *et al.* (1988).

the initial frequency jump. Fig. 6.2, however, showed another transient, amounting to a recovery of 0.7% of the initial frequency jump, with a time constant of 1.6 days. These two exponentials account for the recovery within an r.m.s. error of only 35 μs, corresponding to a distance of only about 10 metres along the equator of the pulsar.

Very few glitches, in any pulsar, have been observed in such detail; none has shown such clear evidence for two distinct exponential recovery components. It is of course possible that many glitches have the two exponential recoveries, but with the available data only one can be distinguished. The other Vela glitches, as set out in Table 6.1, are described in terms of a frequency step with a single exponential recovery; probably the short exponential component was missed because the data points are too far apart.

In Table 6.1 the division between the initial effects of the glitch, and the recovery, may involve some interpretation of the observations; the original references should be consulted for more detail. For example, the factor Q, which is the fraction of the initial step in period that is recovered, can only be determined if the precise date of the step is known. The dates (and modified Julian Days) of most of the earlier events are only known approximately.

6.3 The Crab Pulsar PSR 0531+21

This pulsar is the youngest known pulsar; the period is 33 ms and the rate of change $\dot{P}$ = 36.4 ns/day, giving a characteristic age $\frac{1}{2}P/\dot{P}$ = 1250 years. Long runs of timing data are available, spanning twenty years, during which time the pulsar has made more than 10^{10} rotations.

Although the timing data on the Crab Pulsar has some gaps of several months, all individual observations of pulse arrival times can be correctly numbered, with only occasional ambiguities in the sequence during the longest gaps. The steady increase in period must obviously be allowed for, because it has a surprisingly large effect even on a short run of timing data. Using a linear change in P only, the arrival time τ of the Nth pulse becomes after time t ($\sim NP$)

$$\tau = NP + \frac{1}{2}\frac{\dot{P}}{P}t^2. \tag{6.1}$$

The accumulated correction $\frac{1}{2}(\dot{P}/P)t^2$ becomes one tenth of a period in only one day. The precision of the timing observations is such that the 'age' $\frac{1}{2}P/\dot{P}$ of the Crab Pulsar can be measured in one day to an accuracy of 1%.

Irregularities in pulse timing only appear on a comparatively long time scale. The steps, or 'glitches' in this pulsar mostly appear as changes amounting to only one part in 10^8 of the period as contrasted with one part in 10^6 for the Vela Pulsar. Glitches occurred in the Crab Pulsar in 1969, 1975 and 1986; monitoring was incomplete during the years 1978 to 1981, and another glitch could have occurred unnoticed during that time. The event of August 1986 was observed in great detail (Lyne and Pritchard 1987). Fig. 6.3 shows the timing deviations, calculated by fitting the best values of frequency ν and its first two derivatives to a series of observations prior to the glitch. The recovery is then an exponential with time constant 5.5 days, leaving a steady phase change of 3 milliseconds.

The first glitch, in September 1969, was observed by Boynton *et al.*

(1972) in much less detail. The step increase in frequency was one part in 10^9; there is possibly also an exponential recovery with time constant between 5 and 7 days. The glitch of February 1975, observed by Lohsen (1975) and analysed in detail by Demiansky and Proszynski (1983) shows a step in rotation frequency $\Delta\nu/\nu = 4\times10^{-8}$, and an exponential recovery component with time constant of 10 days. Most notably, there was also a persistent increase in the slowdown rate $\dot\nu$, amounting to $\Delta\dot\nu/\dot\nu = 2\times10^{-4}$. The effect is that the pulsar is now rotating more slowly than it would have without the glitch. Lyne, Pritchard and Smith (1988) have analysed all the available pulse arrival times from 1969 to 1987; they find that the step in $\dot\nu$ was a unique event during the whole of this epoch.

In Section 6.7 we consider the interpretation of glitches in terms of the coupling between the superfluid interior and the crust of a pulsar. An important parameter is found to be the change $\Delta\dot\nu$ in first derivative $\dot\nu$ immediately after the glitch. This was only observed directly after the 1986 event, but values can be inferred for the first two glitches from the estimated exponential recovery terms. In all three cases the initial change $\Delta\dot\nu/\dot\nu$ is of the order of 0.1%. Since the frequency step $\Delta\nu/\nu$ was also comparable for all three, we might reasonably regard the three observed glitches as rather regular occurrences, which suggests that a fourth occurred undetected in 1980 or 1981.

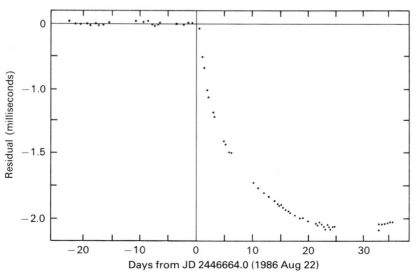

Fig. 6.3. The glitch in the Crab Pulsar, August 1986 (Lyne & Pritchard 1987).

6.4 Timing discontinuities in other pulsars

Table 6.2. *Glitches in other pulsars*

PSR	MJD	Initial jump $\frac{\Delta\nu}{\nu}$	$\frac{\Delta\dot{\nu}}{\dot{\nu}}$	Recovery τ_{days}	$\left(\frac{\Delta\nu}{\nu}\right)_f$	$\left(\frac{\Delta\dot{\nu}}{\dot{\nu}}\right)_f$	Reference
0355+54	46079	5.6×10^{-9}	1.8×10^{-3}				L1987
	46468	4.4×10^{-6}	~0.1	44		6×10^{-3}	
0525+21	42064	1.3×10^{-9}	5×10^{-3}	~143			D1982
	43834	0.3×10^{-9}		<4			
1325−43	43590				1.2×10^{-7}		N1981
1508+55	41740	2.2×10^{-10}			6×10^{-3}		M1974
1641−45	43327				1.9×10^{-7}		M1978
1907+00	42162				6.8×10^{-10}		G1976
2224+65	43070				1.7×10^{-6}		B1982

References: Lyne (1987); Downs (1982); Newton *et al.* (1981); Manchester & Taylor (1974); Manchester *et al.* (1978); Manchester *et al.* (1983); Gullahorn *et al.* (1976); Backus *et al.* (1982).

6.4 Timing discontinuities in other pulsars

Observation of pulse arrival times have been made for most known pulsars over a sufficient length of time to obtain accurate positions and frequency derivatives. The total length of these data strings amounts to some hundreds of years. In this time only a very few discontinuities were observed in any pulsars apart from the Crab and Vela Pulsars. Table 6.2 lists the clearest cases so far observed. For most, only the final change in frequency $\Delta\nu/\nu$ was observed, but for several there were sufficient obser-

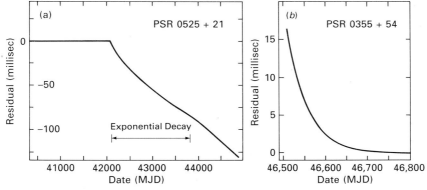

Fig. 6.4. Glitches in two pulsars: (*a*) PSR 0525+21, (*b*) PSR 0355+54. The model in (*a*) results from a fit of exponential recovery (Downs 1982).

vations to follow the transient behaviour. Two examples in older pulsars are shown in Fig. 6.4 (PSR 0525+21 and PSR 0355+54).

The two entries in the Table for PSR 0355+54 indicate the possible complexity of the glitch phenomenon. The first was observed only as a small step in frequency $\Delta \nu/\nu \approx 5.6 \times 10^{-9}$. The second, one year later, was the largest observed in any pulsar, with $\Delta \nu/\nu = 4.4 \times 10^{-6}$, and it was followed by a transient with decay time constant 40 days. There was also a small permanent change in slowdown rate, amounting to $\Delta \dot{\nu}/\dot{\nu} = 0.5\%$. The initial change in $\dot{\nu}$, immediately after the glitch, was about 10%; the precise value depends on the date of the glitch, which is only known to about one month. This step in $\dot{\nu}$ is larger than any previously measured in any pulsar. We refer to this in Section 6.8.

Fig. 6.5. Timing noise (Helfand *et al.* 1980).

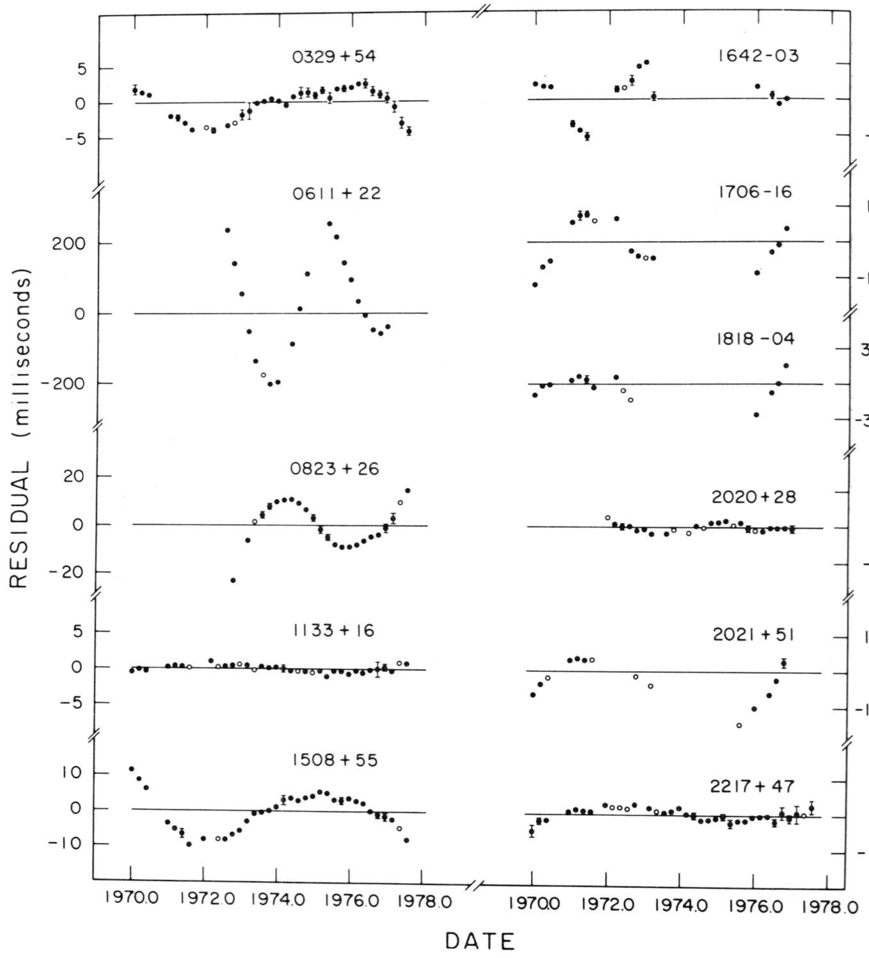

6.4 Timing discontinuities in other pulsars

The glitch in PSR 0525+21 is of note because the period of this pulsar is one of the longest known. This, and the Crab Pulsar, which has a period over a hundred times shorter, produces glitches that are very similar except for the recovery time constants, which differ by a factor of more than 10.

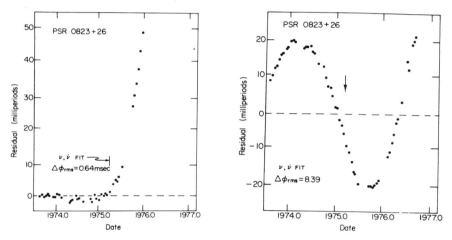

Fig. 6.6. Timing noise in PSR 0823+26. The same data analysed in two different ways: (*a*) using a best fit to the first part of the data (*b*) fitting to the complete run of data (Helfand *et al.* 1980).

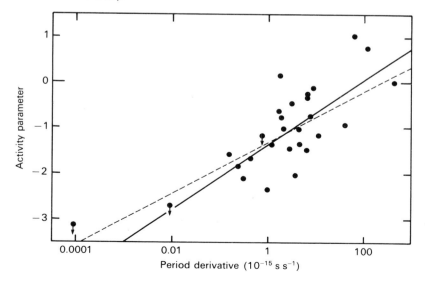

Fig. 6.7. The relation between timing activity and period derivative $\dot{P}$ (Cordes & Downs 1985).

6.5 Timing noise

Pulsars are very good clocks. Allowing for the steady increase in period, they are stable to an accuracy of one part in 10^{11} or more (Cordes and Downs 1985), apart from the rare discontinuities described earlier in this chapter. At this accuracy, however, there are often clearly discernable random irregularities in the periods. These are measured as phase deviations in the rotation of the pulsars, on the assumption that pulse timing is exactly synchronised to pulsar rotation.

These random deviations are known as timing noise, which in the example of Fig. 6.5 shows as residual phase deviation in a sequence of timing observations, after making allowance for the best fit of period and first period derivative. The exact shape depends on the way the fit is made and the length of the sequence under analysis. For example Helfand *et al.* (1980) show that the phase residuals for PSR 0823+26 over a three-year span can be represented as in either (*a*) or (*b*) of Fig. 6.6. In (*a*) the data for the first two years only are used to find best values of v and $\dot{v}$; the remainder of the data then show a rapidly increasing deviation. In (*b*) the fit is made over the whole span of the data; the residuals now take the quite different form of a sinusoid with a period of three years. Evidently neither provides an unambiguous representation of a physical process within the pulsar. Three possibilities must be considered: the fluctuations might be part of a wide spectrum of noise-like components, they might be a quasi-periodic oscillation, or they might indicate an unexpectedly large value of the second differential $\ddot{v}$.

The level of timing noise is characterised by an r.m.s. level over a specified period of observations, usually of order three years, after making a best fit for period and its first derivative. Cordes and Helfand (1980) present results from long series of data on 50 pulsars. For the Crab Pulsar, their analysis gives an r.m.s. deviation of 12 milliseconds; this is taken as a standard, and the other results are presented as an 'Activity Parameter', which is the logarithm of the ratio of the r.m.s. deviation to the value for the Crab Pulsar. There is a good correlation between 'Activity' and the derivative $\dot{v}$, or $\dot{P}$ as shown in Fig. 6.7; for example, the Crab Pulsar, with the highest known value of $\dot{P}$ is very active, while, fortunately, the binary pulsar for PSR 1913+16 and the millisecond pulsar PSR 1937+21 both have very low timing noise, corresponding to very low values of $\dot{P}$. We can conclude that activity is related either to the rate of slowdown, or possibly to age: the young pulsars are almost always the most active ones.

Three different types of noise can, in theory, be distinguished; they are

6.5 Timing noise

Table 6.3. *Random timing noise for the most active pulsars*

PSR	Activity parameter	Noise type
0329+54	−1.1	Frequency
0531+21	0.0	Frequency
0611+22	1.0	Slowdown
0823+26	−0.1	Slowdown
1133+16	−1.9	Phase
1508+55	−0.7	Frequency
1915+13	−1.0	Frequency
1001+31	−0.9	Frequency
2016+28	−1.8	Frequency
2020+28	−1.5	Frequency
2217+47	−1.5	Phase

After Cordes and Helfand 1980.

called respectively phase, frequency, and derivative, or slowdown noise. Each corresponds to a random walk behaviour in a measured parameter, respectively, ϕ, ν and $\dot{\nu}$. Phase noise would result from a random movement of the source of emission over the surface of the pulsar. Frequency noise would result from a random fluctuation in moment of inertia, for example from random changes in ellipticity. Slowdown noise, i.e. noise variations of frequency derivative, would result from random changes in magnetic field or in the structure of the magnetosphere.

Timing noise was strong enough in 11 of the 50 pulsars studied by Cordes and Helfand to allow the distinction to be made. The analysis is simple; the r.m.s. of the deviations from the best fit varies with the time span t in different ways for the three types of noise, increasing respectively as $t^{1/2}$, $t^{3/2}$ and $t^{5/2}$. The result is seen in Table 6.3.

The types of noise seem to progress from phase, for the lowest level of activity, through frequency, which accounts for the majority of pulsars, to slowdown, for the highest level of activity. It is not clear whether separate physical causes are at work; although frequency noise is the most usual case, we must beware of any over-simplified interpretations of discontinuities and timing noise. The problem is compounded by the more recent results for the Crab Pulsar presented by Lyne, Pritchard and Smith (1988). Here a consistent set of timing observations covering a period of five years, and analysed in the conventional way, yielded the timing deviation of Fig. 6.8. It now appears that the best description may be a quasi-periodic

74 Timing irregularities

oscillation. The period, about 20 months, is not influenced by the length of the data string. Unlike the case of PSR 0823+26 in Fig. 6.6 there is no sinusoidal component with a period of about five years, i.e. the length of the data string, which might be an artifact of the analysis.

We will consider in Section 6.9 the possibility that there is an oscillation with a period of 20 months in the superfluid interior of the Crab Pulsar.

6.6 Adjustments in ellipticity: starquakes

The discontinuities and noise-like irregularities in pulsar timing provide a remarkably penetrating means of investigating the interior structure of the pulsars. The sensitivity of these observations is demonstrated by considering the small change ΔR in radius R, which, if the neutron star shrank uniformly, would give an observed speed-up $\Delta \nu/\nu$ of one part in 10^9 (corresponding to the typical discontinuity in the Crab Pulsar). The corresponding change in moment of inertia I is given by

$$\frac{\Delta I}{I} = \frac{2\Delta R}{R} = -\frac{\Delta \nu}{\nu}. \tag{6.2}$$

For a star of uniform density and a radius of 10 km the change would be only 5μm, a remarkably small quantity to be observed in an object at a distance of 2 kpc. A more realistic explanation is in terms of a change in

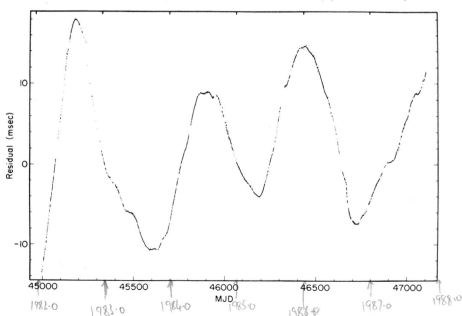

Fig. 6.8. Oscillatory timing residuals in the Crab Pulsar (Lyne *et al.* 1988).

ellipticity ε, accompanied by a sudden cracking of the solid crust to adjust to the new, more spherical shape: again the change $\Delta\varepsilon$ must be small in the observed Crab glitches, since the moment of inertia is $I_0(1 + \varepsilon)$ where I_0 is the value for a spherical body. The equilibrium oblate form of a rotating body has ellipticity ε determined by the ratio of angular kinetic energy to gravitational energy, so that

$$\varepsilon = I\Omega^2 \left(\frac{GM^2}{R}\right)^{-1} \tag{6.3}$$

If, in adjusting to a reduction in Ω, the eccentricity changes by $\Delta\varepsilon$ the strain energy released is proportional to $\mu V(\Delta\varepsilon)^2$, where μ is the shear modulus of the crust, and V its volume. The crust has a very strong crystal structure, with a high value of shear modulus. The energy released in a typical Crab glitch is about 4×10^{39} erg as compared with a total stored gravitational energy of 2×10^{42} erg. A similar glitch could therefore be expected to occur every few years during a lifetime of 10^3 years.

For the Vela Pulsar the results of this calculation are totally different. If the much larger timing discontinuities are to be such 'crust quakes', the ellipticity would have to change by over 3% in a single step, while the energy released would be so great that such events would necessarily be separated by hundreds of years instead of the observed two or three years.

We can conclude that minor glitches, and possibly also the restless behaviour of pulsars observed as timing noise may be accounted for by these adjustments of the crust, but that the major discontinuities certainly cannot.

6.7 Two-component models: the glitch function

A typical glitch in both the Crab and Vela pulsars consists of a step change in angular velocity followed by a partial recovery over a period of time, as in Figs. 6.2 and 6.3. This behaviour is well represented by the expression

$$v(t) = v_0(t) + (\Delta v)_0 [Q\exp(-t/\tau) + 1 - Q], \tag{6.4}$$

which is now known as the 'glitch function' (Baym et al. 1969). Values of τ, the time constant of the exponential recovery, and Q, the fractional recovery of the step in the rotation frequency, are given in Tables 6.1 and 6.2 above.

The form of the glitch function indicates that the pulsar does not rotate as a single rigid body, but that it has two components which are only loosely coupled together. The majority of the moment of inertia is associated with a component that may be thought of initially as the solid crust.

A second component, which is part of the fluid interior, rotates independently but is loosely coupled to the solid crust. The steady slowdown of rotation affects the crust, but leaves the liquid component rotating faster by a fixed differential amount depending on the frictional forces coupling the two components. The discontinuity consists of a sudden increase in coupling between the two components, which speeds up the crust and slows down the liquid. They they decouple again, and differential rotation is slowly re-established.

It is clear from the long decay times τ that the frictional forces are extremely small, which indicates that the liquid is in a superfluid state. This must be part of the neutron fluid, but not all of it, as will be seen shortly. The solid component consists of the crust and all those parts of the fluid that are tightly coupled to it. This must include the electron and proton components of the fluid, which are coupled to the crust by the magnetic field.

Identification of the two components (and indeed of the extra component shown up in the one detailed observation of a Vela glitch) requires a closer look at the dynamics of the fluid. Alpar *et al.* (1981) show that the main bulk of the liquid core is effectively co-rotating with the solid crust; these together form a single component. The separately rotating component is to be identified with that part of the neutron superfluid which co-exists with the crystal lattice in the inner part of the solid crust. This superfluid component can rotate at a different angular velocity from the crust, and the rotational coupling between fluid and crust can change. In the initial step of the glitch, the superfluid suddenly shares its excess angular momentum with the rest of the pulsar. Immediately afterwards, it is completely decoupled from the pulsar rotation, and partial coupling is again restored during the glitch recovery phase. This process of coupling and uncoupling is an unfamiliar concept which we now describe.

6.8 Vorticity in the neutron fluid

We recall that the outer crust of a typical neutron star is about 5 km thick, the total radius being about 15 km. The inner part of the crust is a crystal lattice of neutron rich nuclei, permeated by free neutrons. The liquid interior is composed mostly of neutrons, in equilibrium with a small fraction of electrons and protons. Both neutron regions are superfluid, but in different states of superfluidity. As in the Bardeen, Cooper and Shrieffer (BCS) theory of superconductivity, the attractive force between neutrons with energies close to the Fermi energy leads to pairing, and their energy spectrum then shows a gap of order 1 MeV wide. The superfluid

6.8 Vorticity in the neutron fluid

within the crust involves $^1S_0$ pairing; at the greater densities of the interior the pairing is $^3P_2$, giving an anisotropic superfluid.

Rotation in a superfluid is not a uniform co-rotation of the whole volume; instead it is in the form of vortices. Each vortex carries a quantum of angular momentum, so that the rotation of the superfluid core is measured by the area density of the vortices; for the Crab Pulsar, with angular frequency 200 rad s^{-1}, there are about 2×10^5 vortices per square centimetre. Each vortex is very small; the core of a vortex is of order 10^{-12} cm across. As the pulsar slows down, the rotation of the superfluid can only keep pace by an outward migration of the vortices from the rotation axis.

The rotation of the interior superfluid is tightly coupled to the rotation of the crust. This occurs through the small population of electrons and protons in the neutron fluid. The electrons are coupled to the magnetic field of the crust, and they are also coupled to the vortices through magnetic forces.

A different regime is found in that part of the superfluid which penetrates the inner parts of the crystalline crust. Here the nuclei themselves contain superfluid neutrons, and there is an interaction between the vortices and the nuclei. As pointed out by Anderson and Itoh (1975), the vortices may become pinned to the nuclei, and so be unable to migrate outwards. The angular momentum of this superfluid component is then fixed, and this component then makes no apparent contribution to the moment of inertia of the neutron star. The glitch is understood as the catastrophic breakdown of this pinning, when the angular momentum is suddenly shared with that of the rest of the neutron star, giving the observed speed-up.

The equilibrium situation is not a complete pinning in all regions, but a steady outwards creep of vortices through the crystal lattice. The rate depends on the energetics of the pinning, on the temperature, and on the differential rotation rate between the creep region and the crust. Immediately after the glitch the vortex density is close to that required by the rotation rate of the crust, the pinning is complete, and the angular velocity of the superfluid is therefore fixed. In this condition the rotation of the superfluid is uncoupled from that of the crust. As the crust slows down, the differential rotation increases, and creep is re-established.

We can find the amount of superfluid which is decoupled at the time of the glitch from the change $\Delta\dot{v}$ in slowdown rate immediately after the glitch. If the moments of inertia are I_f and I_t for the fluid and the total, then

$$\left(\frac{\Delta\dot{v}}{\dot{v}}\right)_{t=0} = \frac{I_f}{I_t}$$

where $\dot{v}$ is the slowdown rate in the steady state. The observed values are given in Tables 6.1 and 6.2. Between 0.1% and 10% of the total moment of inertia of the star is decoupled in these glitches.

The tables also show the residual change $(\Delta\dot{v}/\dot{v})_f$ after the transient is complete. This change, which is not always observed, must represent a part of the superfluid that remains firmly pinned to the nuclei of the crust without re-establishing the steady outwards creep of the equilibrium state. It is, of course, quite possible that this component also will re-couple on a much longer time scale.

6.9 Vortex lattice oscillation

Tkachenko pointed out in 1966 that the vortices in a rotating body of superfluid must be arranged in a regular lattice structure; furthermore, any displacement of this structure would lead to a wave motion consisting of a transverse, or shear wave. In 1970, Ruderman suggested that some apparent oscillations with about a four-month period in the timing of the Crab Pulsar might be due to Tkachenko oscillations in the neutron superfluid of the interior. The idea was abandoned when it became clear that the timing oscillations were noise-like rather than periodic, as discussed in Section 6.5. Now that we have the five-year run of timing residuals shown in Fig. 6.8, the oscillations may be real, although with the much longer period of 20 months.

The Tkachenko theory gives a velocity for the shear wave in the vortex lattice that depends on the density of the vortices, i.e. on the rate of rotation. For a neutron star interior, with radius 10 km, theory predicts an oscillation period of $20P^{1/2}$ months, where P is the rotation period in seconds. As Ruderman (1970) pointed out, this predicted an oscillation period of four months for the Crab Pulsar, agreeing with the apparent oscillation as seen in 1970. If the 20-month oscillation is to be interpreted in this way, the mode of oscillation must be different; for example, there may be a solid core at the centre of the neutron star. Oscillations might then be a relative movement between the core and the crust, rather than a pure lattice oscillation in the fluid. If the oscillation is a decaying transient, as appears from Fig. 6.8, then it may be stimulated by a glitch.

An alternative explanation of the oscillation is that the 20-month period is the period of precession of the rotation axis, similar to the precession of the X-ray source Her X–1 or to the Chandler wobble of the Earth's axis. Precession of the crust would couple to the superfluid interior, giving rise to the observed oscillation in rotation rate.

Finally, we may speculate that at least part of the timing noise observed

6.9 Vortex lattice oscillation

in other pulsars is due to internal oscillations. The oscillation period would usually be several years, possibly following the $P^{1/2}$ dependence of Tkachenko waves, so that the limited data span of the present observations would not provide any clear distinction between an oscillation, random noise, or a large value of $\ddot{P}$ which might follow a glitch. It will be well worth while continuing the timing observations for some years to come.

7

The young pulsars

Although it is now accepted that most neutron stars are born in supernova explosions, only three of the known pulsars (the Crab Pulsar, the Vela Pulsar, and PSR 1509−58) are clearly associated with visible supernova remnants. This is, of course, entirely consistent with the difference between the lifetimes of a typical pulsar (about 10^{6-7} years) and a supernova remnant (about 10^{4-5} years). Furthermore, these three pulsars are obviously young, as seen from their small periods P and large period derivatives $\dot{P}$. Their characteristic ages, $\tfrac{1}{2}P/\dot{P}$, are in sharp contrast with those of the other short-period pulsars, the short-period binaries and the millisecond pulsars (Chapter 9); these have very small period derivatives, are much older, and may have had a different origin or different evolutionary history.

The three young pulsars are to be regarded as the youngest members of the general population of pulsars. Their predicted progress through the P–$\dot{P}$ diagram (Fig. 5.2) will bring them into the general population within 10^6 years, and it is only their youthful characteristic of a large energy output that makes them especially interesting. The rate of loss of rotational energy is

$$\dot{E} = I\omega\dot{\omega} = 4\pi^2 I\dot{P}P^{-3}. \tag{7.1}$$

Since the moment of inertia I is almost independent of the detailed model adopted for the neutron star, and not much dependent on its mass, the total energy output is available for any pulsar directly from the measured values of P and $\dot{P}$.

The three young pulsars all radiate in the radio, optical, X-ray or gamma-ray regimes, covering an astonishingly wide range of the electromagnetic spectrum. The Crab Nebula, surrounding the Crab Pulsar, depends totally on the pulsar for its energy supply; the nebulae surrounding the Vela Pulsar and PSR 1509−58 radiate less spectacularly, and are less dependent on their central pulsars.

These rapidly evolving pulsars have a large derivative in rotation rate, $\dot{\nu} = -\dot{P}/P^2$, which is easily observed, and the second derivative $\ddot{\nu}$ can also be measured. The large steps in rotation rate, or glitches, which occur in the Vela Pulsar make it difficult to assign an average value of $\ddot{\nu}$ to this pulsar. Timing observations for the other two have given values of $\ddot{\nu}$ consistent with theory; for the Crab Pulsar a value of the third derivative $\dddot{\nu}$ has been found, which again is consistent with the theory in which the slowdown follows the law

$$\dot{\nu} = -k\,\nu^n \tag{7.2}$$

where n is the braking index (Chapter 5).

The short timescale of evolution in these young pulsars, and the very energetic radiation that they produce, make their study particularly rewarding. Some of the interesting differences between them emerge in the accounts of the different ways in which they were discovered.

7.1 The Crab Pulsar PSR 0531−21
7.1.1 *Discovery*

We referred in Chapter 1 to the hunt for a pulsar in the Crab Nebula, both by optical and radio techniques. The motivation came from several indications of some unknown but very energetic process located in the centre of the nebula, and probably centred on an unusual star, the south-west member of a pair of bright stars. The whole of the nebula, described in more detail in Chapter 9, generated radio, optical and X-ray radiation at such a rate that its brightness should fade within a few years. This showed that it had an energy supply quite apart from the energy of the supernova explosion observed 900 years ago. This radiation was shown by Shklovsky to be synchrotron radiation (see Chapter 14) from electrons with relativistic energies moving in a magnetic field. The optical radiation from the central star was also not understood for many years. It was very intense, blue in colour and without spectral lines in absorption or emission. There was also known to be a very compact radio source near the centre of the nebula, with an unusual radio spectrum. This radio source had been isolated from the general radio radiation from the nebula by Cambridge radio astronomers (Hewish and Okoye 1964), who had observed that it scintillated, implying that it had a very small angular diameter.

The final clue, unnoticed by most observers, was provided by a theorist, Pacini, who suggested in 1967 that a rapidly rotating and highly magnetised neutron star would radiate an electromagnetic wave at its rotation frequency with sufficient energy to provide for the requirements of a nebula like the Crab Nebula.

These were sufficient clues to initiate an exciting hunt. The discovery came in the radio band, but by that time there already were recordings, not yet fully analysed, in both optical and X-rays which clearly contained the evidence of pulsations at a period of 33 milliseconds. There was, however, no reason at the time to expect so short a period, and the search analysis becomes more difficult at shorter periods. The radio discovery at Greenbank (Staelin and Reifenstein 1968) was in fact a detection of individual radio pulses and not of a periodicity. It is a peculiarity of the Crab Pulsar that it emits occasional very large pulses, the so-called giant radio pulses. The detecting system was designed to detect such individual pulses, which were to be recognised by the dispersion in their arrival times in a number of adjacent radio frequency channels. It was only after some of these individual pulses had been found that the Fourier analysis technique at Arecibo was extended down to the very short period of 33 milliseconds, and a different detection technique, which took advantage of the precise periodicity, could be used.

It is interesting to reflect that without the chance detection of the giant pulses, neither the Greenbank search nor any of the general radio surveys made over the next 16 years would have detected a pulsar of so short a periodicity. The chances would otherwise have favoured the optical searches, which were already working with sufficient time resolution. Given the vital clue of the actual period, it was not long before the optical pulses were discovered, and the nature of the blue central star was revealed.

7.1.2 *Pulse profile*

The Crab Pulsar emits pulses over more than 60 octaves of the electromagnetic spectrum. At the lowest observable frequency, 10 MHz, the periodicity is lost because of smearing by multipath propagation in the interstellar medium (Chapter 17), but there is no indication that this is an actual lower frequency limit to the emission. At the highest energy, there is a clear detection in gamma rays of 500 GeV, and a possible detection at 2000 GeV; the limit of emission may have been reached, but it may also be that detection systems are too insensitive and the photons too few for the measurement of the spectrum to be extended further.

Over the whole of this range the pulse profile is remarkably and astonishingly similar. Fig. 7.1 shows the development of the two component profile over the radio range of frequencies from 74 MHz to 430 MHz. Two features distinguish the lower part of this range. Pulse smearing has a progressively larger effect at frequencies below 400 MHz;

7.1 The Crab Pulsar PSR 0531−21

the length of the smearing varies with radio frequency approximately as v^{-4}. Also, an additional component can be seen immediately before the main pulse. This is known as the precursor and it is seen only at radio wavelengths.

Fig. 7.2 shows the comparison between optical, X-ray and gamma-ray pulse profiles. There is a gap in the spectrum between short radio wavelengths ($\sim$ 10 cm) and infrared ($\sim$ 3.5 μm), partly on account of atmospheric absorption and partly due to a dip in the pulsar spectrum (see below); in the ultraviolet the pulse profile has not yet been measured, although there is no apparent dip in the mean intensity in this part of the spectrum.

The similarities are obvious. For example, the spacing between the two peaks is the same at all wavelengths, and the shape of the peaks shows no effect such as flattening due to self-absorption at any part of the spectrum. This can only mean that the same population of particles, located in the same region of the magnetosphere, is responsible for the radiation over the whole spectrum. The only exception is the radio precursor; in Chapter 15 we attribute the precursor to radiation from a different location within the magnetosphere.

Various minor differences can be pointed out, such as:

(i) at X-ray and low gamma-ray energies there is a more obvious

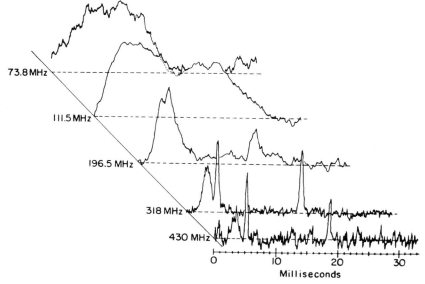

Fig. 7.1. Pulse profile for the Crab Pulsar over the radio frequency range 74 MHz to 430 MHz (Drake 1971).

84 *The young pulsars*

bridge between the main pulse and the interpulse, and the interpulse increases relative to the main pulse;

(ii) the peaks are narrower at radio wavelengths, apart from the effects of interstellar scattering.

The pulse profile can be measured in greatest detail at optical wavelengths. Fig. 7.3 shows the profile obtained by Jones *et al.* (1981), in which light was detected over the whole of the pulse period. The intensity

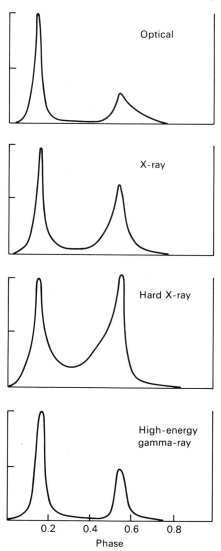

Fig. 7.2. Optical, X-ray and gamma-ray pulse profiles for the Crab Pulsar. (For references see Wilson & Fishman 1983.)

7.1 The Crab Pulsar PSR 0531−21

falls to 2% of the peak between the pulses, and to 0.9% at the lowest point. This light curve clearly resolved the sharp peak of the main pulse, which can be fitted by a Gaussian curve with width 3°.5, which is similar to the beamwidth in many radio pulsars. The optical profile is very stable when expressed in terms of angular rotation; a comparison of profiles from 1970, 1977 and 1985 shows no detectable evolution (apart from a stretching of the whole profile due to the period increase), even though this interval of time represents 1.5% of the age of the pulsar.

7.1.3 Spectrum

The radio frequency spectrum of the Crab Pulsar can be extended down to below 20 MHz despite the overwhelming effect of pulse broadening at low frequencies. Interferometer measurements by Soviet astronomers extending down to 16.7 MHz (Bovkun 1979) have revealed a small diameter continuous source in the Crab Nebula, which may safely be assumed to be the pulsar. Fig. 7.4 shows the mean radio intensity from 16.7 MHz to 1660 MHz. The spectrum is steep, with an index of −2.7 over most of its length. The radio precursor has a very much steeper spectrum than the other radio pulse components. It is almost undetectable above 610 MHz, where its height is one tenth that of the main pulse, but below 150 MHz it dominates the pulse profile. The spectral index of the precursor is therefore about −5 at high frequencies, but it is presumably about −3 at low frequencies.

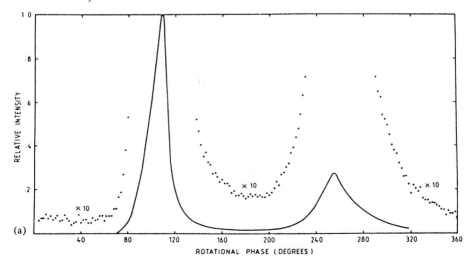

Fig. 7.3. Optical pulse profile for the Crab Pulsar. The expanded curve (dotted) shows that the intensity falls to 1% immediately before the main peak (Jones *et al.* 1981).

The optical, X-ray and gamma-ray spectra are continuous, and may be seen with the radio spectrum in Fig. 7.4. The continuous, high-energy spectrum may extend to energies above 1 TeV, i.e. above 2×10^{27} Hz; it is, however, difficult to check that the pulse profile is the same here as at lower energies. The logarithmic plot of Fig. 7.4 shows a power-law spectrum with index varying from -1.1 at gamma-ray energies, through -0.7 at X-ray energies to zero in the optical range; it then changes rapidly to $+2$ in the long infrared. The broken portions of the plot are not yet observed, but the extrapolation seems well justified.

It is instructive to re-plot the spectrum to show the emitted power as a function of frequency. Fig. 7.5 shows $\log \nu I_\nu$ against $\log \nu$; it is seen that the emitted power is concentrated in the short X-ray spectral region.

The gap between the radio and infrared is evidently more than a result of the technical difficulty of observing at wavelengths around 100 μm. The turn-down in spectrum in the infrared, established in measurements by Middleditch *et al.* (1983), takes the spectrum to more than a factor of ten below the extrapolated high energy spectrum. This infrared turndown is not accompanied by any dramatic change in the pulse profile, as might be

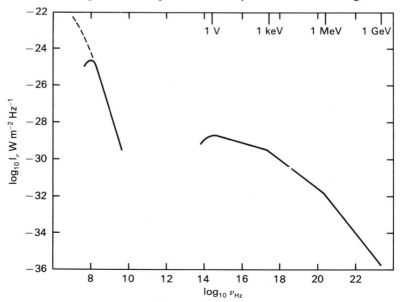

Fig. 7.4. Intensity I_ν of the pulsed radiation from the Crab Pulsar on a logarithmic plot covering more than 16 decades (53 octaves) of the electromagnetic spectrum. The spectrum below 100 MHz (dashed line) is the unpulsed source, presumed to be the pulsar whose pulses are lost by interstellar scattering.

7.1 The Crab Pulsar PSR 0531−21

expected if it were to be accounted for by the effects of saturation or self-absorption (Chapter 16).

7.1.4 Polarisation

At radio wavelengths the precursor pulse is highly linearly polarised, with a constant position angle. The integrated profiles of the main pulse and interpulse are about 20% polarised, and the position angle is nearly constant through the two pulses. Precise measurements are difficult, because of the weakness of the signal at short wavelengths and the effects of interstellar multipath propagation at long wavelengths.

More precise measurements are available at optical wavelengths (Smith *et al.* 1988). High degrees of linear polarisation are seen throughout the pulse profile with rapid swings of position angle at the two peaks (Fig. 14.3). This is a strong indication that the light is synchrotron or curvature radiation from high-energy electrons flowing along magnetic field lines. We explore this interpretation in Chapter 14.

7.1.5 Intensity fluctuations – giant pulses

There is a very marked contrast between the extreme variability of the Crab Pulsar at radio wavelengths, where fluctuations are observed on timescales ranging from microseconds to years, and its very regular behaviour at optical and X-ray wavelengths, where no fluctuations are

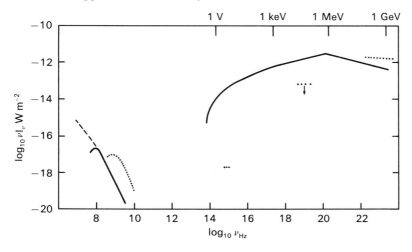

Fig. 7.5. The distribution of emitted power: the spectrum of Fig. 7.4 plotted as $\log \nu I_\nu$ versus $\log \nu$. The major part of the power is in high-energy X-rays. The dotted curves show the more limited observations of the Vela Pulsar, including the upper limit on the X-ray flux.

detectable either from pulse to pulse or over periods of years. At the highest gamma-ray energies, where detection involves the air-shower technique, there is a suggestion that the radiation is confined to bursts 15 minutes long (Gibson *et al.* 1982); this sporadic behaviour is, however, difficult to confirm because of the large statistical fluctuations in the small number of photons at such high energies. It seems reasonable at present to associate major fluctuations only with wavelengths longer than the infrared; this is an indication that the radiation mechanism is fundamentally different below and above the gap in the observed spectrum.

A test of the variability of optical pulses was made by Hegyi, Novick and Thaddeus (1971). The intensity of each pulse was recorded as a photon count N, and the two averaged quantities $\langle N \rangle^2$ and $\langle N^2 \rangle$ were compared. Over a typical averaging time of five minutes, no difference was observed in the ratio of these two quantities at any part of the main pulse profile. This excluded the possibility of pulse-to-pulse fluctuations similar to those observed at radio wavelengths.

The radio pulse-to-pulse variations in the Crab Pulsar are among the largest for any pulsar. The 'giant' pulses, which led to its discovery, are the extremes of a continuous distribution of pulse intensity (Fig. 7.6, Argyle

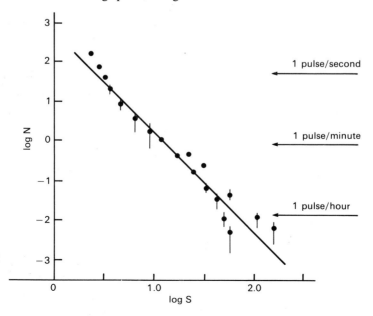

Fig. 7.6. Pulse height distribution for the Crab Pulsar (Argyle & Gower 1972). N is the number of pulses per minute above pulse strength S, measured in units of the average pulse strength.

7.1 The Crab Pulsar PSR 0531−21

and Gower 1972), in which a giant with ten times the mean intensity occurs about once per 10^3 pulses, i.e. once per 30 seconds on average. Hankins and Rickett (1975) showed that these giant pulses are very short, typically lasting 100 microseconds or less, so that their peak intensities are very large indeed, corresponding to brightness temperatures of 10^{31} K or more. They occur mainly close to the time of the main peak in the pulse profile: about 7% appear near the centre of the interpulse, and none at the time of the precursor or at other phases of the pulse.

The giant pulses are only observed at radio frequencies below 500 MHz. Heiles and Rankin (1971) found little correlation between their intensities recorded simultaneously at 74, 111 and 318 MHz indicating that the phenomenon is narrowband.

7.1.6 Intensity fluctuations – long-term scintillation

As with all other pulsars, the radio emission from the Crab Pulsar is variable on a wide range of time scales. Until Rickett, Coles and Bourgois (1983) pointed out that scintillation in the interstellar medium could be responsible for long-term variations in apparent intensity of small-diameter radio sources, all fluctuations with periods of more than a few minutes were regarded as an unknown, slow, intrinsic process in the pulsar. Rankin, Payne and Campbell (1974) found a characteristic fluctuation time of 30 days at 430 MHz. Other observations at 196 MHz gave a longer characteristic time, while a sequence at 610 MHz observed at Jodrell Bank gave a shorter timescale (Fig. 7.7). The interpretation is clear: these are the long-term refractive focussing effects in which the timescale increases with wavelength, as contrasted with the short-term diffractive scintillation in which the timescale decreases with wavelength (Chapter 17).

This is, however, not the whole story. Lyne and Thorne (1975) found that the intensity at 408 MHz increased by a factor of four over a period of three months at the end of 1974. This may yet be attributed to refractive focussing, although the long timescale does not fit the theory at present. A long series of observations on several different frequencies simultaneously would be needed to resolve this issue.

There was no optical variability corresponding to this major radio variation.

7.1.7 Variable radio pulse broadening

The multipath propagation effect, which broadens the radio pulses at long wavelengths (Fig. 7.1), is also variable. The effect of pulse broadening is measured by a characteristic time, which varies approximately as v^{-4}; the effect is therefore quoted as a scattering parameter τ_p, where the characteristic time is $v^{-4}\tau_p$. This parameter is normally about 20×10^5 MHz^4 for the Crab Pulsar. In the first half of 1974 this parameter increased dramatically to 800×10^5 s MHz^4 (Lyne and Thorne 1975), and subsequently decreased to normal over a period of several months. The broadening function also showed substantial variations of shape during this period.

A study of the broadening under normal conditions by Rankin and Counselman (1973) had already shown that there were probably two components to the broadening. It now became clear that one component was constant, and referred to the major part of the transmission path, while the other, variable component must relate to the interior of the Crab Nebula. These variations were reasonably attributed to an electron cloud or filament crossing the line of sight.

There are also variations in the Dispersion Measure (DM) of the Crab Pulsar. Some care must be taken in distinguishing these from the effects of pulse broadening, since both constitute pulse delays which decrease with frequency. However, dispersion delay varies as v^{-2}, while the broadening varies approximately as v^{-4}, allowing the variation of DM to be distinguished in observations at the shorter radio wavelengths. Fig. 7.8 shows variations in DM observed in a series of timing observations at Arecibo

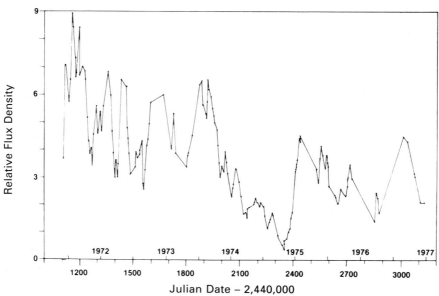

Fig. 7.7. Slow refractive scintillation of the Crab Pulsar (Rankin *et al.* 1988).

7.1 The Crab Pulsar PSR 0531−21

and Jodrell Bank. Again, the variations are attributed to fluctuations in the transmission path within the Crab Nebula, apart from the obvious events in June each year when the pulses are delayed by their passage through the solar corona.

7.1.8 Crab pulse timing

Several extensive series of timing observations have been made both in the optical and radio regimes. A summary and analysis up to April 1979 is given by Demianski and Proszynski (1983), and a five-year series of radio observations at Jodrell Bank has been analysed by Lyne, Pritchard and Smith (1988). Radio observations can be made more nearly continuously, since they are unaffected by daylight; the pulse arrival times must, however, be compensated for interstellar dispersion, which is found to vary appreciably.

If a sequence of several years of observed pulse arrival times, corrected to the barycentre of the Solar System (Chapter 5), is fitted to a simple polynomial, allowing for a best fit in rotational frequency v and its two derivatives $\dot{v}$ and $\ddot{v}$, the residual derivations in pulse arrival times are usually contained within one pulse period. These deviations, which may include a glitch discontinuity as well as a quasi-periodic wander, were discussed in Chapter 6; they are believed to be related to a variable coupling between the separate components inside the neutron star.

A different analysis, which investigates the slowdown rate of the pulsar over the whole span of observations, presents values of $\dot{v}$ separately obtained from each month of observation. The result is plotted in Fig. 7.9(a), where it appears that the slowdown rate is changing linearly with time. The expanded plot in Fig. 7.9(b) is obtained by subtracting this linear rate, i.e. a single value of $\ddot{v}$. Two effects can now be seen: a step change in 1975, and a slow curvature. The latter can be removed by adding a single value of $\dddot{v}$, giving the residuals of Fig. 7.9(c).

Fig. 7.8. Variations of dispersion measure DM over a period of 15 years.

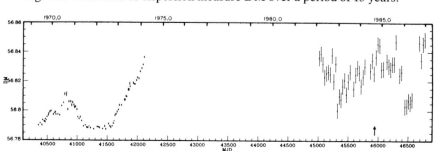

92   *The young pulsars*

The braking index *n* (see Section 5.5) is related to these derivatives by

$$n = \frac{\nu\ddot{\nu}}{\dot{\nu}^2} \tag{7.2}$$

and

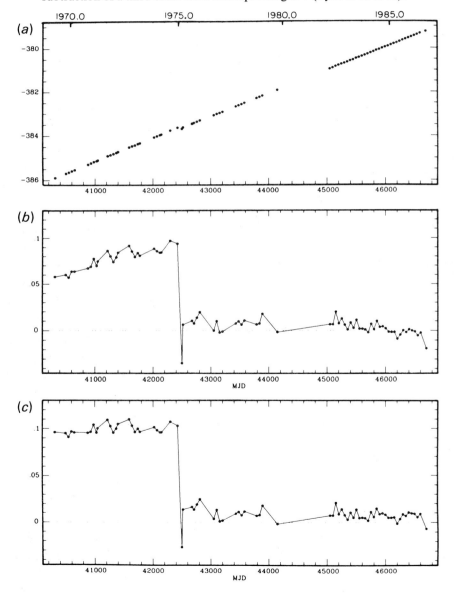

Fig. 7.9. (*a*) Slowdown rate $\dot{\nu}$ for the Crab Pulsar. (*b*) The same after subtraction of a linear rate of change, i.e. a fixed value of $\ddot{\nu}$. (*c*) The residuals after further subtraction of a third order term corresponding to $\dddot{\nu}$ (Lyne *et al.* 1988).

7.2 The Vela Pulsar PSR 0833−45

$$n(2n-1) = \frac{\dddot{v} \cdot v^2}{\dot{v}^3} \tag{7.3}$$

The measured values of the derivatives give a value $n = 2.509 \pm 0.001$ from equation 7.2 and $n = 2.5 \pm 0.3$ from equation 7.3. The slowdown of the Crab Pulsar does indeed follow a power law, but not precisely the law expected from pure magnetic dipole radiation where the braking index is $n=3$. An explanation of the difference must involve a deeper understanding of the energy loss in the outflow of energetic particles (Chapter 5).

7.2 The Vela Pulsar PSR 0833−45

7.2.1 *Discovery*

The Vela Pulsar was discovered in a general search for pulsars in the southern sky, using a technique which was designed to detect single pulses. Individual pulses from the pulsar were so strong that they immediately showed their periodicity of 89 ms on the pen chart recordings. This was the shortest pulsar period then known.

The discoverers, Large, Vaughan and Mills (1968), immediately noted that the pulsar is located close to the centre of the Vela nebula, a supernova remnant 4° or 5° in diameter. This nebula was itself already known to be a radio source, and their first publication unequivocally identified the pulsar as a rotating neutron star located at the origin of the supernova explosion. The association of the pulsar with this supernova remnant may not be as certain as at first appeared, since the geometrical centre of the remnant seems to be far removed from the pulsar (Bignami and Caraveo 1988).

Unlike the history of discoveries for the Crab Pulsar, there was no detection of pulses from the Vela Pulsar in X-rays, and the optical pulses were so faint that they were only discovered after a prolonged effort (Wallace *et al.* 1977). There is, however, a strong gamma-ray pulse, with a different pulse profile (Buccheri 1981).

7.2.2 *Pulse profile*

Fig. 7.10 shows the integrated pulse profiles for the Vela Pulsar in the radio, optical, and gamma-ray spectral regions. The contrast with the Crab Pulsar is at first sight dismaying. The closest resemblance is in gamma-rays (Kanbach *et al.* 1980), where two components are separated by 0.4 of the period, and the space between is partly filled. But the two optical components (Manchester *et al.* 1980), shown correctly phased in Fig. 7.10, occur between the gamma-ray pulses; if they correspond at all,

94   *The young pulsars*

they are associated with the continuous gamma-ray emission between the two peaks. Furthermore, the radio pulse occurs before the first gamma-ray peak: the only hope of relating it to the gamma-ray and light curves is to identify it with the precursor seen in the radio pulse from the Crab Pulsar. This identification turns out to provide the key to the geometry of the various emitting regions (Chapter 14).

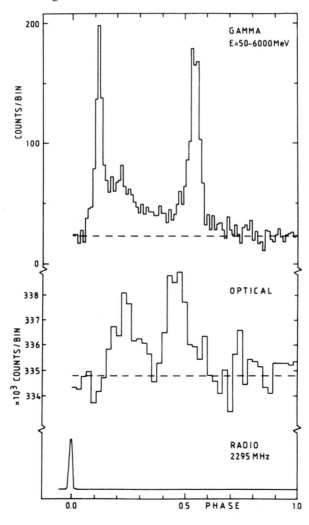

Fig. 7.10. Pulse profiles of the Vela Pulsar in the radio, optical and gamma-ray regions.

7.2 The Vela Pulsar PSR 0833−45

7.2.3 The spectrum

The radio spectrum of the Vela Pulsar, shown in Fig. 7.5 is similar to those of many other pulsars (Chapter 16), with a spectral index of −2.4 and a low-frequency turnover at about 600 MHz.

The Vela Pulsar is one of the strongest radio pulsars, the strongest gamma ray source in the sky, and at the same time one of the faintest detectable optical objects and too weak to be detected in X-rays. The spectrum in Fig. 7.5 plotted as vI_v versus log v, is evidently very incomplete. A comparison with the Crab Pulsar suggests that the spectrum of the Vela Pulsar shows a similar dip between the radio and optical, but that the peak energy is radiated in the gamma-ray region rather than optical and X-ray.

In the high-energy gamma-ray region the spectral index α, defined as $I_v \propto v^\alpha$, is approximately −0.9. It is not appropriate to assign power-law indices to the rest of the spectrum, as the variations in pulse profile show that the various components have very different spectra.

7.2.4 Polarisation of the integrated radio profile

The Vela Pulsar provides the classical, and the first, example of a highly linearly-polarised, integrated, pulse profile, in which the plane of polarisation swings smoothly through a large angle, in this case about 120°. This swing was interpreted by Radhakrishnan *et al.* (1969) in terms of a rotating vector source related to the field structure near a magnetic pole. The polarisation behaves similarly over a wide range of radio frequencies.

We note that the high polarisation and steep spectrum of the radio pulse from the Vela Pulsar are also seen in the precursor radio pulse from the Crab Pulsar. Our geometrical analysis in Chapter 14 identifies both as typical radio pulses, as from the majority of pulsars, and considers the high-energy pulses from the Crab and Vela Pulsars as a separate phenomenon.

7.2.5 Vela pulse timing

In the young pulsars, the slowdown in rotation rate is easily detected, and the discovery paper of the Vela Pulsar included a value of $\dot{P}$. Measurements both in Australia and in USA then showed a major step (Chapter 6), in which the period decreased suddenly by 200 ns, as compared with the steady daily increase of 10.7 ns.

Five such events were recorded up to 1983. Fig. 7.11 shows that these might be considered as minor discontinuities in the steady slowdown, but they represent nevertheless a catastrophic change in the interior of the neutron star. If, for example, they were to be accounted for by a change in radius, giving a corresponding change in moment of inertia, there would

96   *The young pulsars*

have to be a change of 1 cm at each step. The accepted explanation is in terms of locking and unlocking of superfluid motion within the pulsar (Chapter 6).

Each event consists of a step and a recovery. The unpredictability of glitch events, and the long interval between them, makes it difficult to monitor an event in detail. The observations of October 1981 missed the actual glitch by less than a day. They showed that there were two stages of recovery, with time constants of 1.6 days and 233 days (from McCulloch *et al.* 1983).

The series of glitches, with long recovery times and variable steps in period, make it impossible to measure a steady value of second derivative $\ddot{P}$ for the Vela Pulsar.

7.3   PSR 1509−58

The third example of the young pulsars was discovered first as an X-ray source by the UHURU satellite, and later by the Einstein satellite as a pulsating source with period 150 ms. The X-ray pulse is broad, occupying 25% of the period between half power points. It resembles pulses from the binary X-ray sources, in which the pulse shape is determined by the changing aspect of a hot spot at a magnetic pole. The X-ray spectrum appears to be thermal, with brightness temperature around 10^7 K. It was not obvious that such a source would be a radio pulsar.

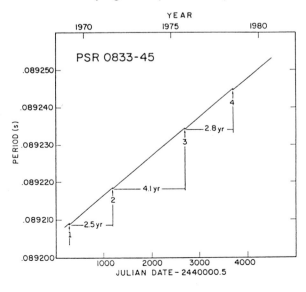

Fig. 7.11. The slowdown of the Vela Pulsar, from late 1968 to mid-1980, showing four major glitches (Downs 1981).

7.3 PSR 1509−58

There was, however, a radio source already known at this position. It is the source MSH 15−52, or G320.4−1.2 (from its galactic coordinates), which is associated with the visible nebula RCW 89, believed to be a supernova remnant. The X-ray source is located near the centre of this radio source, which is about ½° across (Caswell *et al.* 1981), but there is no obvious discrete radio source at this location to be seen on the radio map (Fig. 7.12). A search for a radio signal at the known periodicity was, however, successful. Manchester and Durdin (1984) have now measured the period derivative $\dot{P}$ and the second derivative $\ddot{P}$, showing that the slowdown follows expectation for a young pulsar, with characteristic age −1500 years, and braking index 2.8 ± 0.2.

We therefore have a supernova remnant, which seems to be at least 10 000 years old, containing at its centre a radio pulsar only 1500 years old. The pulsar is behaving normally in the radio domain, but has the peculiarity of pulsed X-ray emission, which apparently has an origin different from that of the radio pulses. The radio pulsar is weak and has a large dispersion measure ($DM = 235 \pm 25$ pc cm^{-3}); it probably would not have been found in a radio search without the prior X-ray discovery.

Fig. 7.12. The radio and X-ray sources near PSR 1509−58 (Seward *et al.* 1983). The shaded area is the extended radiosource, believed to be a supernova remnant; the X-ray source is outlined by a broken line. PSR 1509−58, marked as a dot, is both a radio and an X-ray pulsar.

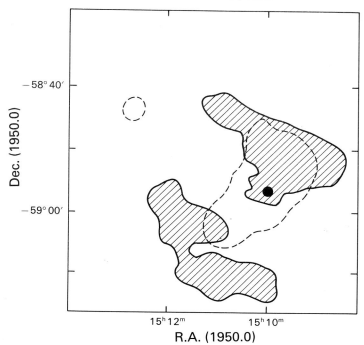

Table 7.1. *Young pulsars*

PSR	P_{sec}	$\dot{P}(10^{-15})$	Age (years)
0531+21	0.033	420	1 200
1509−58	0.150	1540	1 500
0833−45	0.089	120	11 000
1800−21	0.133	134	16 000
1737−30	0.607	466	20 000
1823−13	0.101	72	22 000
1930+22	0.144	58	40 000
2334+61	0.495	192	40 000
1727−47	0.830	164	80 000
0611+22	0.335	60	89 000
1916+14	1.181	211	89 000
1951+32	0.039	6	104 000
0656+14	0.385	54	112 000

From Braun *et al.* 1989.

The discrepancy in age suggests that the nebula is the remnant of an earlier explosion, perhaps of a former binary companion, which led by rapid mass transfer to the second supernova explosion which was the origin of the pulsar.

The apparently thermal X-radiation is harder to understand. The pulsar timing results are remarkably smooth, and the possibility of accretion sufficient to heat the polar regions and produce X-rays is therefore unlikely.

7.4 Other young pulsars

Another young pulsar has been discovered in a supernova remnant (SNR 0540–69.3) in the Large Magellanic Cloud (Seward *et al.* 1984, Middleditch and Pennypacker 1985). This has been detected in X-rays and optically, with a period of 50.2 ms, and a large slow-down rate as for the Crab Pulsar. The age appears to be about 660 years.

Further examples of young pulsars can only be expected from organised and extensive searches. A major difficulty is that they are most likely to exist close to the galactic plane, where they are particularly difficult to detect because of dispersion and scattering by interstellar ionisation. This forces the searches to be made on a higher radio frequency, where the beam width of the telescope is smaller, so that the sky is searched more slowly. A successful survey of this type was made by Clifton and Lyne (1986), as described in Chapter 3. Among the 40 pulsars discovered were three that have characteristic ages of about 20 000 years. They are younger

7.4 Other young pulsars

than any others apart from the four pulsars we have discussed in this chapter, all of which have visible supernova remnants. The characteristics of the 13 youngest pulsars are given in Table 7.1.

Apart from the Crab and Vela Pulsars, no supernova remnants can be seen in these positions either in radio or optically; they might have disappeared in the lifetime of the pulsars, but in any case they would be hard to see optically at such distances in the galactic plane. An exciting possibility is to detect gamma rays from these young pulsars, using the next generation of space-based gamma-ray telescopes.

8

The galactic population of pulsars

8.1 The surveys

A glance through the catalogue of the known pulsars shows at once that they are mostly found in the Milky Way. They must therefore be young galactic objects, and it might be assumed that their distribution through the Galaxy is similar to that of young massive stars and supernovae. Although this is nearly correct, it can only be established by reading the catalogue in conjunction with a description of the surveys in which the pulsars were found; many of these surveys in fact concentrated on the plane of the Galaxy, giving an obvious bias to the catalogue, while others show considerable variations of sensitivity over the sky.

The first surveys to cover large areas of the sky were comparatively insensitive, and necessarily gave rather meagre evidence. For example, Large and Vaughan (1971) found only 29 pulsars in 7 steradians of the southern sky. Nevertheless this catalogue, combined with a northern hemisphere catalogue covering low galactic latitudes (Davies, Lyne and Seiradakis 1972, 1973) showed that there were at least 10^5 active pulsars in the Galaxy. This number, together with the pulsar ages known from measurements of slowdown rates, suggested that the birthrate of pulsars might be as high as one every ten years (within a factor of three).

There are now over 400 known pulsars, most of which have been discovered in four surveys carried out at Arecibo, Jodrell Bank, Molonglo and Greenbank, all at frequencies near 400 MHz (Hulse and Taylor 1974, 1975; Davies, Lyne and Seiradakis 1972, 1973; Large and Vaughan 1971; Manchester *et al.* 1978; Damashek *et al.* 1978). The entire sky has now been surveyed to a reasonably well calibrated flux density limit, and the results can be used to determine the distribution of pulsars both in luminosity and spatially throughout most of the Galaxy. The analysis that follows is largely due to Lyne, Manchester and Taylor (1985).

8.2 The observed distribution

Fig. 8.1 shows the distribution on the sky of the 316 pulsars detected in the four surveys. The plot is in galactic coordinates, and the concentration of pulsars along the galactic plane is clear. The depth of coverage is, however, far from uniform over the sky. The low density of pulsars towards the centre of the Galaxy is due to the reduced receiver sensitivities caused by the high galactic background noise and interstellar scattering in this direction (Section 8.7). The concentration of pulsars observed near galactic longitude 50° is due to a deep Arecibo survey, which covered only this region of sky.

The distances of these pulsars are found primarily from their dispersion measures, using a model of the galactic distribution of electrons as described in Chapter 4. Given the distances, we can now plot the distribution of observed pulsars projected onto the galactic plane (Fig. 8.2), and the distribution in $|z|$, the distance from the plane (Fig. 8.3). As expected, the inverse square law has a very large effect on the apparent distribution in distance from the Sun in the galactic plane. The true population must now be evaluated by taking into account this and the other selection effects peculiar to the various surveys.

8.3 The derived luminosity and spatial distributions

For each pulsar we can obtain an estimate of the radio luminosity $L = S_{400}d^2$ from the observed flux density S_{400} at 400 MHz and at the

Fig. 8.1. Distribution on the sky of 316 pulsars. These constitute a sample over the whole sky to a well-defined flux limit.

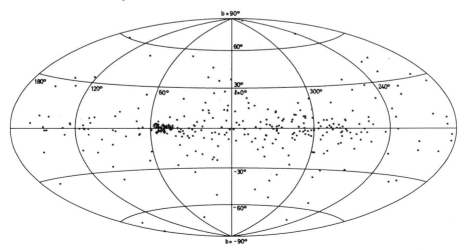

distance d. Each survey has a lower limit in luminosity, which depends on distance and which may also vary with position on the sky. Taking into account the parameters of each survey, the observed distribution in luminosity L, galactic radius R, and distance z from the plane are then

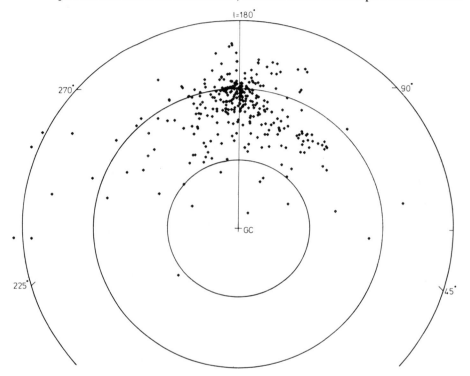

Fig. 8.2. The positions of the 316 pulsars in the uniform sample, projected onto the plane of the Galaxy. The galactic centre is at the centre of the diagram. The pulsars are clustered round the Sun, at a distance of about 8 kpc from the centre.

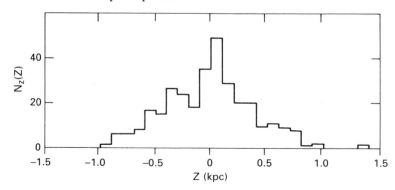

Fig. 8.3. The distribution in $|z|$, the distance from the galactic plane, of the uniform sample of pulsars.

8.3 The derived luminosity and spatial distributions

transformed into a model of the actual space distribution. The distribution in the Galaxy is assumed to be axially symmetric.

Three separated and independent density distributions $\varrho_R(R)$, $\varrho_z(z)$, $\phi(L)$ can now be constructed. (We later discuss the possibility that the distributions in z and L may be partly correlated: the present analysis uses insufficient data to investigate this separately.) The derived distributions in L, R and z are shown in Figs. 8.4, 8.5 and 8.6 along with the distributions as observed.

The z-distribution is approximately an exponential with scale height 400 pc. This is a much wider distribution than that of most galactic Population I objects, which have scale heights of order 50–100 pc; we discuss this later in terms of the high observed pulsar velocities. The radial distribution ϱ_R shows an increasing density towards the centre of the Galaxy, as expected for supernova remnants; the apparent lack of pulsars for $R <$ 4 kpc may be due to an observational difficulty to which we will return in Section 8.7. The surface density, projected on to the galactic plane at the radial distance of the Sun, is 70 ± 15 observable pulsars per square kiloparsec.

The distribution in luminosity is shown in equal intervals of log L. Over more than three decades the distribution fits the empirical law

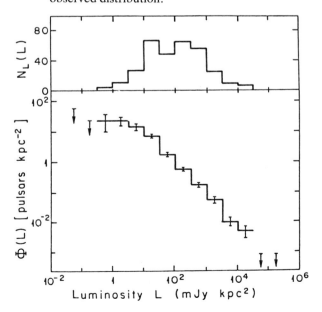

Fig. 8.4. The derived distribution $\varphi(L)$ in luminosity, compared with the observed distribution.

$$N(L)\,dL \propto L^{-2}\,dL.$$

Fortunately, the sensitivity of the surveys is sufficient to show that the population falls well below the extrapolation of this law to low luminosities, since the integration that gives the total population would otherwise be divergent for low luminosities. Even so, we have virtually no information on pulsars with luminosities below 0.3 mJy kpc^2.

The result of integrating over all three distributions gives a total of 70 000 observable pulsars with luminosity $L > 0.3$ mJy kpc^2 in the Galaxy, within a statistical accuracy of 25%. A further uncertainty of about 20% should be included to take account of the uncertainty in the distance scale.

The total number of active pulsars is larger than the total number observable by a factor depending on the angular width of the radio beam. The analysis of beamwidth in Chapter 12 shows that this factor must be approximately 5; we deduce that the total number of active pulsars is $(3.5 \pm 1.5) \times 10^5$ for $L > 0.3$ mJy kpc^2.

It should be noted that, although only about 4% of the observed population have luminosities between 0.3 and 3 mJy kpc^2, they represent between 80 and 90 per cent of the galactic population. The pulsars observed from the Earth represent mostly a small minority of very high luminosity objects, with a mean luminosity about 100 times that of most of the galactic population. Out of the large population of low luminosity pulsars, only a few that are close to the Earth are actually observable.

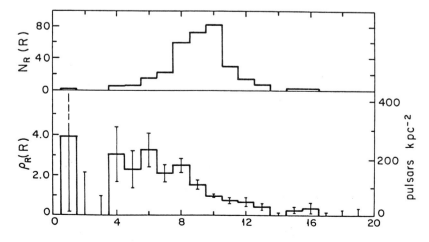

Fig. 8.5. The derived distribution $\varrho(R)$ in radial distance from the galactic centre, compared with the observed distribution.

8.4 Pulsar velocities and ages

The scale height of 400 parsecs in z-distance, as contrasted with much smaller values for massive stars and supernova remnants, was explained by Gunn and Ostriker (1970); they suggested that pulsars are runaway stars that have moved large distances after their birth. A large proper motion for PSR 1133+16 was found from timing observations by Manchester, Taylor and Van (1974), and direct interferometer measurements by Lyne, Anderson and Salter (1981) showed that transverse velocities between 100 and 200 km s^{-1} were usual. Fig. 8.7 shows the velocities of pulsars relative to the galactic plane. With only two exceptions, the velocities are directed away from the plane, entirely confirming the Gunn and Ostriker suggestion.

If, as seems reasonable, these pulsars were born close to the galactic plane, their present ages can be found by extrapolating their velocities back to the plane. These ages can then be compared with the 'characteristic' ages $\tau_c = \frac{1}{2} P/\dot{P}$, obtained from their present rate of slowdown. Fig. 8.8 shows this comparison. The simple conclusion is that the agreement is satisfactory for young objects, but for older ones the real ages are less than the characteristic ages, the discrepancy increasing with age. This again agrees with a suggestion by Gunn and Ostriker, that the dipole magnetic

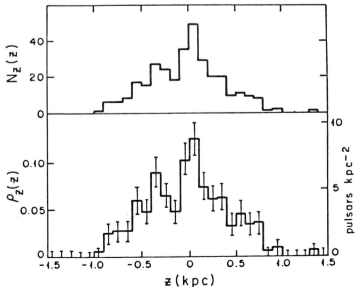

Fig. 8.6. The derived distribution $\varrho(z)$ in distance $|z|$ from the galactic plane, compared with the observed distribution.

106 *The galactic population of pulsars*

Fig. 8.7. The measured velocities of pulsars, showing the general movement away from the galactic plane. The filled circles show the present position (in z-distance and galactic longitude), and the tails show tracks of their motion in the last million years.

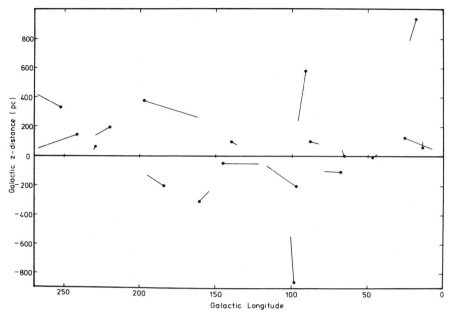

Fig. 8.8. The characteristic ages of pulsars compared with their 'dynamic' ages derived from their velocities and distances from the galactic plane. The three curves show the tracks expected for magnetic field decay on three time scales.

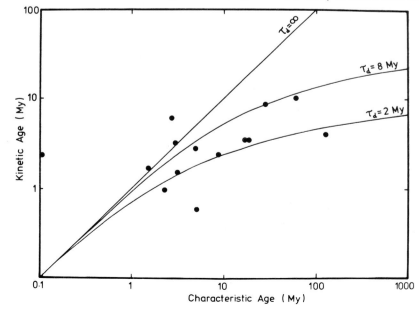

6.4 Pulsar velocities and ages

field decays with age. This causes the slowdown rate, $\dot{P}$, to decrease on a time scale t_B, with a corresponding increase in the measured characteristic age $\tau_c = \frac{1}{2}P/\dot{P}$. The velocity data of Fig. 8.8 suggest that t_B is between 2 and 8 million years (My).

A further probable consequence of this decay is a decay of luminosity. This would imply a relation between luminosity and z-distance, contrary to the assumption made in separating the population functions in section 8.3. A consistent picture is obtained if the luminosity is proportional to the square of the field B, so that the luminosity decays with time constant $\frac{1}{2}t_B$. A direct test of this is shown in Fig. 8.9, where the measured luminosity is plotted against characteristic age. The curve showing luminosity proportional to $P\dot{P}$, i.e. to B^2, is derived for a luminosity decay time of 4.6 million years, i.e. $t_B \approx 9$ My.

The decay of the magnetic field means that pulsars do not traverse the P–$\dot{P}$ diagram in a straight line, as shown in the logarithmic plot of Fig. 8.10. As the field decays, so $\dot{P}$ falls below this line, whose slope is -1. This affects the distribution of the characteristic ages of pulsars in a flux-limited sample. Fig. 8.11 shows that the observed distribution fits the calculated distribution for $t_B = 9$ My very well indeed.

Fig. 8.9. The decay of luminosity with characteristic age. The curve shows the expected fall in luminosity for a magnetic field decay time of 9 million years, both for luminosity proportional to $P\dot{P}$, or B^2, and to $\dot{P}$, or B^2P. The observed values of luminosity are for groups of pulsars with various characteristic ages.

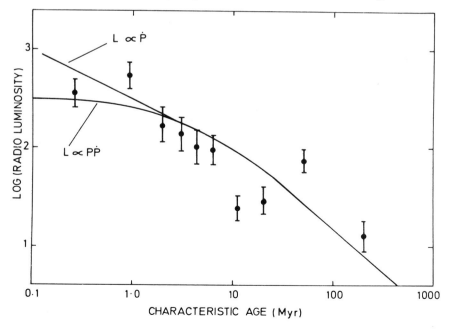

8.5 The pulsar birthrate

The lifetime of the main population of pulsars, of the order of 10 million years, is so small compared with the age of the Galaxy that we can assume a constant population of pulsars, subject only to statistical fluctuations. The luminosity function in Fig. 8.4 therefore represents a steady rate, in which individual pulsars are moving from left to right as their luminosities decay. The individual bins in this plot cover a range $\sqrt{10}$ in luminosity. We know, therefore, that pulsars move to the right by one bin in 4.6 My. The maximum number N of pulsars in one bin is found near the minimum of the luminosity function, where $N \approx 22$ kpc^{-2}. The birth-

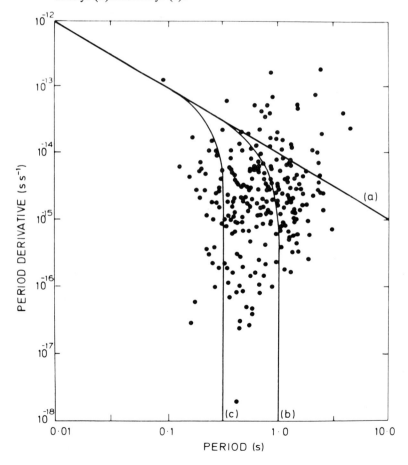

Fig. 8.10. The evolution of pulsars in the P–$\dot{P}$ diagram. For a constant dipolar magnetic field, the evolution follows the line (a) with slope -1; a fall below this line indicates a decay in the field. The curved tracks are for field decay times of 10 Myr (b) and 1 Myr (c).

8.5 The pulsar birthrate

rate near the Sun must therefore be 4 kpc^{-3} My^{-1}, within a factor of two. Integrating through the Galaxy, and multiplying by five for the beaming factor, we obtain a required birthrate of one every 46 years. Combining all uncertainties we can state that the birthrate lies in the range of one every 30 years to one every 120 years. Blaauw (1985) points out that this is a high rate in relation to the evolution of the known population of stars; for example, the evolution of all stars with mass greater than 9 M$_\odot$ would provide a supernova rate of one per 50 years.

Looking back over the earlier estimates, which gave considerably higher pulsar birthrates, we see two reasons for the difference. It was at first supposed that the decay of luminosity was sudden rather than exponential. This of itself does not have a serious effect, but it has a more significant consequence. The observed pulsars are, as we have seen, predominantly young and very luminous. The lifetime used in the earlier calculations was, therefore, too short, giving an underestimate of the age of the population as a whole and an overestimate of the birthrate in the past.

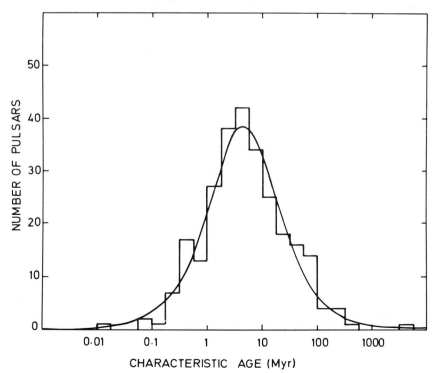

Fig. 8.11. The observed distribution of characteristic ages of pulsars compared with the theoretical distribution for a field decay time of 9 million years (assuming that the luminosity is proportional to B^2).

8.6 The population of millisecond pulsars

The surveys that have been used in determining the pulsar population were not suitable for detecting pulsars with periods of only a few milliseconds. Since the characteristic ages of these fast pulsars are typically much longer than the ages of the normal population, it is to be expected that a considerable population of fast pulsars might exist undetected. Attempts to find such a population have generally failed, and we may conclude that fast pulsars do not change our estimate of total population by more than 10%. If they are born, or reborn, in a different process (Chapter 10), their rate of birth is very much lower than the creation of normal pulsars.

8.7 The galactic centre region

The noise level in a radio telescope looking towards the galactic plane is increased by the high level of the background radiation. This partly accounts for the poor statistics seen at low R in Fig. 8.5. There is now believed to be another factor, related to interstellar scattering rather than to background noise. Many pulsars seen in the plane near the centre suffer from pulse broadening as well as high dispersion (Chapter 17). A survey at 400 MHz (Stokes *et al.* 1985, Dewey *et al.* 1985) specifically designed to detect short-period pulsars in the galactic plane failed to detect any new ones, and it was thought possible that this might be due to a complete smearing out of the pulses. The test of this hypothesis is to observe at a higher frequency, since the length of pulse smearing varies inversely as the fourth power of the frequency.

A recent survey carried out at Jodrell Bank, at 1400 MHz, of the central regions of the galactic plane (Clifton and Lyne 1986) showed that this hypothesis is correct. In an area defined by $0 < l < 100°, 0 < |b| < 1°.0$, this survey discovered 40 new pulsars, none of which had been found in the contemporary 400 MHz survey.

The effect of this on the total population is, however, not very large as only a comparatively small region of the plane was effectively lost to observation in the four main surveys. The existence of enhanced interstellar scattering in these central regions has, of course, a great interest in its own right.

8.8 The origin in supernovae

The birthrate, one per 30–120 years, is reasonably compatible with recent estimates of the rates of supernovae from historical and extragalactic evidence, and with estimates derived from observations of super-

8.8 The origin in supernovae

nova remnants (Chapter 9). The direct association of the three young pulsars (Chapter 7) with supernova remnants encourages us to think that all normal pulsars originate in supernova events. Given the observed high velocities of pulsars, the population distribution corresponds well with that of supernova progenitors.

The calculated rate of birth of pulsars does, however, approximately equal the supernova rate. It supernovae are solely responsible for pulsar production, then most supernovae must produce a pulsar. Many supernova remnants are known that do not contain observable pulsars. Most are hollow shells, showing no sign of excitation by a young central pulsar. It may be, however, that some contain a pulsar that is not seen because of the beaming effect; others are in any case too distant for an average pulsar to be detected within them.

From the evidence as it stands, we may still question whether all pulsars are born in supernovae, leaving open the possibility that some are born in rather less violent and less obvious events, as discussed in Chapter 9.

9

Supernovae

9.1 The nature of supernovae

The obvious association between the Crab Pulsar and the remains of the supernova explosion of AD 1054 leads naturally to the suggestion that all pulsars originate in supernova explosions, and even to the speculation that all supernovae might produce neutron stars, which could become pulsars. This turns out to be an over-simplification and it is necessary to explore the nature of supernovae in some detail before their relation to pulsars can be pursued.

In 1920 Lundmark pointed out that the nova observed by Hartwig in 1885 in the constellation of Andromeda was probably within the Andromeda Nebula itself, and hence very distant and very bright (see a centenary review by de Vaucouleurs and Corwin (1985)). He showed that there were many cases of these extremely powerful novae, and he was the first to associate the Crab Nebula with the Chinese records of the bright star that appeared in AD 1054. The physical significance of these enormous outbursts was appreciated by Baade and Zwicky, who first used the word 'supernovae' in their publication of 1934. They made four very remarkable deductions from the observations:

   (i) the total energy released was in the range 10^{51} to 10^{55} ergs;
   (ii) the remnant could form a neutron star;
   (iii) cosmic rays could have their origin in supernovae;
   (iv) supernova explosions could give rise to expanding shells of ionised gas.

These speculative deductions have been amply justified by later work; only the connection with cosmic rays remains to be firmly established. It is astonishing to reflect that the entirely new concept of neutron stars was correctly included in the list, even though it was a concept that remained unfamiliar to most astronomers up to, and even after, the time of the discovery of pulsars.

9.1 The nature of supernovae

Zwicky devoted much effort to the systematic search for supernovae, using photographic techniques involving tedious comparisons of Schmidt plates from different epochs. Statistical inferences can now be made from the several hundred discoveries made so far; far better statistics are, however, expected from some automated search techniques now in progress. On the basis of present statistics it is reasonable to divide the observed characteristics into two types only, although Zwicky suggested several further sub-divisions. The observed differences are mainly in the spectra: Type I spectra contain emission lines from many heavy elements, while Type II spectra are predominantly from hydrogen. The differences are related to the mass of the progenitor stars: Type II originate from more massive stars, which have an extensive envelope of gas whose radiation obscures the underlying radiation from the exploding outer parts of the collapsing star.

In both types the rise to maximum light occurs very rapidly, usually within a few days (Fig. 9.1). At maximum, the light is mainly in a continuum whose spectrum corresponds to a temperature of about 15 000 K.

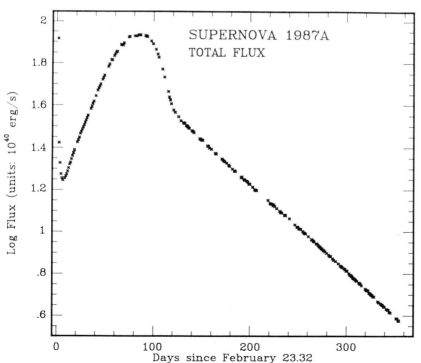

Fig. 9.1. Bolometric light curve of SN 1987A (South Africa Astronomical Observatory).

The total luminosity is about 10^{43} erg s^{-1}, mainly in the visible. A combination of these data gives a radius of about 10^{15} cm at this stage, i.e. about 10^4 R$_\odot$. The subsequent decay can be observed for two years or more, following a monotonic decay which is approximately exponential with a time constant of between 50 and 100 days. The total energy emitted in the visible range is of order 10^{48} ergs.

The spectrum of a Type I supernova usually shows an absorption dip near 6150 Å, attributed to Si II. This enables a measurement of expansion velocity to be made, which is typically around 11 000 km s^{-1}. Bands later develop in the spectrum, which are attributed to Fe$^+$ and Fe^{++}. The differences between the two types appear in the spectrum after maximum, when the Type II supernova develops hydrogen emission lines in the Balmer series but does not develop the Fe$^+$ and Fe^{++} emission bands.

9.2 Stellar collapse

The source of energy for a supernova explosion is the gravitational energy released when a star collapses. The collapse is a result of the failure of the central energy supply, which throughout the normal life of a star provides thermal pressure to balance gravitation. The 'Main Sequence' consists of stars in which the balance is maintained through a steady nuclear burning of hydrogen to helium. Depletion of the hydrogen fuel occurs more rapidly for more massive stars: the hydrogen burning phase for a star with the mass of the Sun (1 M$_\odot$) lasts about 10^{10} years, while a star with 10 M$_\odot$ lasts only 10^6 years.

Helium burning follows the depletion of hydrogen in the core. The central temperature rises to 2×10^8 K, allowing helium nuclei to fuse and form oxygen and carbon. The hot core has a sufficiently high density that the gravitational force at its surface balances the large outward thermal pressure. But the outer parts of the star expand, forming an opaque and comparatively cool envelope. The star is now a red giant; the proportion of mass in the envelope, and the diameter of the star, both increase with the total mass. The core itself is now a compact star, whose evolution proceeds independently of the giant envelope surrounding it.

Progressive stages of nuclear burning, eventually reaching the end-point of iron, beyond which no further energy is available from fusion, can all be taking place at the same time in the core of a red giant, in a succession of shells. The energy available decreases at each stage, and each stage accordingly lasts a shorter time. Collapse follows inevitably, but the form of the collapse depends on the mass of the core. Within a small range of mass, roughly from 1.9 M$_\odot$ to 2.5 M$_\odot$, the core will collapse to form a

9.2 Stellar collapse

neutron star, losing about 1 $M_\odot$ in the explosion that follows the collapse. Cores with a mass smaller than 1.9 $M_\odot$ may explode rather than collapse, because in these small cores the carbon burning phase occurs catastrophically, in a 'carbon flash' explosion, and the star is disrupted. If the mass of the core is greater than 1.9 $M_\odot$, which is the core size expected from the evolution of a main sequence star with mass about 8 $M_\odot$, the star will collapse to form a neutron star or a black hole soon after the start of carbon burning. A black hole will result from the collapse of a core with mass greater than about 2.5 $M_\odot$.

We may summarise the discussion so far as follows:
  (i) stars whose masses lie within a limited range evolve to produce cores that can collapse into neutron stars;
 (ii) when the nuclear fuel in the core is exhausted, it will collapse under its own gravitation; this collapse is the origin of the energy of the supernova explosion;
(iii) the collapse to nuclear density, which takes place in a few seconds, is followed by a rebound, in which part of the core and all the outer parts of the star are blown away. The visible supernova results from this exploded material and from the excitation of the surrounding cloud by the radiation released in the collapse and rebound;
(iv) the core may disintegrate, or collapse to a black hole, or form a degenerate star, according to its mass, which itself is determined by the mass of its parent star (apart from the case of binary stars, which we discuss later).

The masses of the progenitor stars whose evolution produces condensed cores following these various evolutionary tracks are only known approximately. Collapse to form a neutron star, via a Type II supernova, occurs for stars with total mass between 8 $M_\odot$ and 15 $M_\odot$. From 2 to 8 $M_\odot$ a star may become a Type I supernova, leaving no neutron star. Above 15 $M_\odot$ the star will become a Type II supernova, but the core may either collapse to a black hole or disintegrate due to another process, the photodisintegration of iron. The rate of occurrence of these three events depends on the distribution of the stellar population with mass (the initial mass function); the population falls rapidly with increasing mass, so that neutron stars are more likely to originate from stars with 8 $M_\odot$ than with 15 $M_\odot$. On the other hand, stars with smaller masses evolve more slowly, so that the relative numbers reaching the three categories of collapse must depend on the age of the galaxy, or the part of the galaxy under consideration. This is borne out by observations of the locations of the two types: Type I is found in all types of galaxy, while Type II is only found in spiral arms, where there is a large and rapidly evolving population of massive stars (Maza and van den Bergh 1976).

9.3 Luminosity decay

Most of the energy of a supernova collapse is radiated away as a very high flux of neutrinos. Since these have very little interaction with other matter they can pass through and exert almost no outward pressure on the collapsing material, so that the collapse is unimpeded and is complete in a fraction of a second. The burst of neutrinos from the supernova SN 1987a, in the Large Magellanic Cloud, was observed on Earth some hours before the expanding cloud of the supernova became visible.

The spectrum of the expanding cloud of a Type I supernova contains traces of heavy elements from the disrupted core, while in Type II we see only the hydrogen that formed the outer envelope. We remarked earlier on the surprising similarity of the light curves of these two types. In both cases it is the events inside the core that determine the rate of release of energy; after the initial explosion the main source of energy is the exponential radio-active decay of nickel to iron. It happens that ^{56}Ni is an easily formed isotope, because it has atomic number Z equal to the neutron number N, and Z is divisible by 4. It can therefore be synthesized directly from ^{4}He, and it has the highest binding energy per nucleon for $Z = N$. It decays according to the sequence:

$$^{56}\text{Ni} \xrightarrow{6.1 \text{ days}} {}^{56}\text{Co} \xrightarrow{77 \text{ days}} {}^{56}\text{Fe}$$

The binding energy is released mainly as gamma rays and neutrinos, both of which can transfer energy to the expanding cloud. This is the source of the luminosity of both Type I and II. We now see that the decay of luminosity follows the 77 day time constant of the decay of ^{56}Co.

The peak luminosity depends on the total mass of ^{56}Ni. This lies in a small range, because the mass of the collapsing core lies in a much smaller range than the mass of the progenitor. The small range of peak luminosity among the observed Type I and II supernovae is therefore explicable. We will, however, be speculating later that there may be other forms of stellar collapse that lead to the formation of neutron stars, but which are not bright, easily observed events.

9.4 Supernovae in binary systems

It is remarkable that the neutron stars associated with X-ray sources (apart from the Crab Pulsar) are all members of binary systems,

9.4 Supernovae in binary systems

while only seven out of the first 450 pulsars to be discovered were found to be in binary systems. The comparative rarity of binary pulsars is accounted for by the disruption of binary systems at the time of the supernova explosion of one of its members, while the X-ray binaries provide a demonstration of a mass transfer between the two members that can profoundly affect the evolution of both.

Binary stars are common among all types of star, and particularly among the early-type stars that are expected to become Type II supernova. About 70% of O and B stars are in binary systems and, of these, about 70% have binary periods shorter than 10 days (Batten 1967). Mass transfer between the two members has left its mark in many of these systems; of the two components of Algol, for example, the more massive is the least evolved star, although this must have been originally less massive than its partner. In these systems the rate of mass transfer, and the stage of evolution at which it occurs, depend not only on the mass of the two components but also on their separation.

Van den Heuvel and De Loore (1973) gave an example of this interactive evolution for a close binary consisting of stars with masses 25 $M_\odot$ and 8 $M_\odot$, with an orbital period of 4 days. Fig. 9.2 shows the sequence of events. The heavier (primary) star naturally evolves more rapidly, reaching the giant stage before the lighter star has appreciably changed. The star can then fill its 'Roche lobe', which is the volume inside which material can orbit round the single star. Mass is then transferred rapidly to the secondary star, and their masses practically interchange. The primary loses all its outer hydrogen-burning layers, leaving only a helium core, in a short-lived stage known as a Wolf-Rayet binary. This core then evolves rapidly, and explodes, leaving a neutron star or a white dwarf. The enlarged secondary star now evolves more rapidly, eventually in its turn filling its Roche lobe and transferring mass back to the primary. Since the primary is a collapsed object, there is a large release of energy as the material falls on it. The system is now a massive X-ray binary, as described in Chapter 11. The final stage comes when the secondary explodes as a supernova, leaving a neutron star, which may become a pulsar. The mass loss in this explosion may be sufficient for the binary system to be completely disrupted, and the pulsar will then be solitary.

Disruption of a binary generally occurs if more than half of the total mass is lost in the explosion of one of its members (Boersma 1961). The analysis is complicated if the binary pair are very close, when there may be effects of the collision of the expanding supernova shell with the companion star (Colgate 1970). A comprehensive analysis by McCluskey and

118 Supernovae

Kondo (1971) shows that disruption is only certainly avoided if the original mass of the exploding star is less than 0.2 of that of its companion, or if less than 0.2 of its mass is expelled. Most Type I supernovae are expected to remain bound, since they have a comparatively small mass, while Type II should usually disrupt, leaving an expanding shell of gas and two high-velocity stars, one probably a neutron star. Blaauw (1961) suggested that some stars observed to have high velocities, such as U Geminorum, could have originated in this way.

9.5   Quiet collapse

The suspicion that massive supernova explosions are not sufficiently frequent to account for the rate at which pulsars are born leads

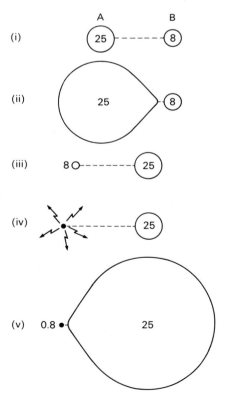

Fig. 9.2. Mass transfer and evolution in a binary system. (i) In the binary system the more massive star A evolves first. (ii) Star A has now expanded, and transfers mass to the lighter star. (iii) Star A now collapses to a small helium star of 8 solar masses. (iv) Star A explodes, leaving a neutron star of 0.8 solar masses. The neutron star may be observable as a pulsar. (v) Star B now evolves, expands, and transfers mass to the neutron star, which becomes an X-ray source.

9.5 Quiet collapse

to a search for other less spectacular origins. The only reasonable direction to look must be towards a way in which a white dwarf can collapse to form a neutron star without expelling a large envelope or exciting a red giant shell. This might occur if a white dwarf in a binary system, with mass about 1.5 $M_\odot$, accreted material from its companion, and collapsed without exploding (Whelan and Iben 1973).

Such accretion-induced collapse may already have been observed. There has been for some years a network of spacecraft orbiting the Earth, equipped with gamma-ray detectors intended to monitor nuclear explosions. On several occasions, an impulse of gamma-rays has indeed been detected (Fig. 9.3; Klebesadel *et al.* 1973), but the origin, as determined by the relative time of arrival at several different spacecraft, has been found to be outside the Solar System. The most outstanding of these

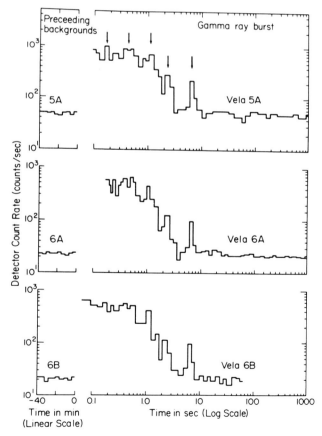

Fig. 9.3. A gamma-ray burst recorded simultaneously by three Vela satellites (1970 August 12; Cline *et al.* 1982).

bursts, on 1979 March 5, was detected by no fewer than nine spacecraft, and it was located within an accuracy of 2 arcminutes: it apparently came from the vicinity of a nebula in the Large Magellanic Cloud (Cline *et al.* 1982).

The proposed explanation (Baan 1982) is that the burst represents the heating of the surface of a newly formed neutron star to a temperature of about 10^9 K as matter falls freely on to it immediately after the collapse. It seems unlikely that this is the only explanation of the gamma-ray bursts, as they are now being detected at a rate of more than one per week.

Accretion-induced collapse of a white dwarf may be comparatively common in close binary systems with lower mass than the example of Fig. 9.2. If, for example, the two masses are 5 $M_\odot$ and 2 $M_\odot$, the mass transfer occurs while the more massive star is approaching the stage of helium burning. Practically the whole of the extended envelope of the star is transferred to its companion, leaving a core which then follows the evolutionary track of a solitary star. If this leads to collapse to a white dwarf, there may eventually be accretion in the reverse direction as the companion evolves. This is probably the origin of the low-mass X-ray binaries, and it may lead to the collapse of the white dwarf into a neutron star, forming a rapidly rotating millisecond pulsar.

9.6 Frequency of occurrence of supernovae

The occurrence of supernovae is such a rare event in any individual galaxy that a measurement of the average frequency of occurrence requires continued observation of a large number of galaxies. Inevitably, the observations are far from continuous and some supernovae will be missed. The extent of this incompleteness is demonstrated by the statistical work of Katgert and Oort (1967), who revised the previous estimate for the average frequency in our Galaxy from one per 450 years (Minkowski 1964) to one per 40 years. A more recent estimate by van den Bergh *et al.* (1987) is one per 28 years.

The evidence from our own Galaxy is less significant than that from other galaxies, since the assessment of completeness requires some guesswork about the thoroughness of ancient observers. No supernovae have been observed since 1572 (Tycho) and 1604 (Kepler), but others may have occurred in obscured regions of the Galaxy. The prime example is the intense radio source Cassiopeia A, which must have exploded about the year 1700. There is good evidence that this was a Type II supernova. It was not reported from any observatory, even though Cas A is circumpolar in the northern hemisphere. The visible remnant, discovered after the source of the radio emission had been located with sufficient accuracy, is heavily obscured, to the extent of 10 visual magnitudes. The supernova explosion would then hardly bring the star into the range of naked eye visibility.

The radio remains of other supernova explosions can be distinguished from other discrete galactic radio sources by their size, shape, and spectrum. Over 100 supernova remnants (SNR) have been catalogued by Ilovaisky and Lequeux (1972). Woltjer (1972) has estimated the lifetime of the remnants and, after allowing for the incompleteness of the catalogues, he estimates that they were formed at a rate between one per 60 years and one per 45 years. The typical age is about 5×10^4 years. These supernova remnants are distributed uniformly over a galactic disc with radius $R = 8$ kpc, with a surface density falling by a factor 10 at $R = 12$ kpc. Within $R = 6$ kpc the z-distribution is close to an exponential with scale height 90 pc. The age and distribution of these remnants agree with an origin in Type II supernovae, which in turn are associated with young Population I stars.

The most remarkable feature of the observed rate of occurrence of supernovae is that the rates for Type I and II supernovae are very similar, despite the fact that they appear to be distinct categories originating in different types of star. Furthermore, the total energy release for the Type II is much greater than for Type I, as demonstrated by the kinetic energy of the expanding remnants; the optical emission is not, however, very different, so that the chances of observation are similar. The ratio of visible energy to kinetic energy may be 10^3 times larger for Type I than for Type II.

9.7 Associations between pulsars and supernovae

As discussed in Chapter 7, only four or five out of more than 450 known pulsars appear to be associated directly with supernova remnants. Looking for pulsars within or close to the hundred or more supernova remnants has not so far yielded any other associations.

The essential parameter in this discussion is the age of the pulsar. If this considerably exceeds the lifetime of the visible supernova remnant, which is of the order of 10^5 years, then no association can be expected. If the supernova remnant is no longer expanding, having encountered a sufficient mass of interstellar material to slow it down from an expansion velocity say 10 000 km s^{-1} to 10 km s^{-1}, then the pulsar velocity may have taken it outside the nebula. The observed associations confirm this view. The Crab Pulsar (age 10^3 years) is near the centre of the Crab Nebula. The Vela Pulsar (age 10^4 years) is within a less well defined supernova remnant.

There is little doubt that most other pulsars have ages in excess of 10^6 years, so that not only have they left the neighbourhood of the supernova explosion; the remnant itself will have disappeared from view.

9.8 The Crab Nebula

As a nebula, the Crab was first observed in 1731 by John Bevis, an English physicist and amateur astronomer. It took first place in the catalogue of nebulae compiled by Charles Messier in 1758, where it appeared as the nebula M1. The name 'Crab Nebula' was given to it about a hundred years later, when better telescopes revealed its tentacle-like structure. The present-day interest in the Crab Nebula dates mainly from the work by Baade in 1942, when he presented observations of its detailed structure and suggested that a prominent star near the centre of the nebula might be related to its origin. Baade already knew that the nebula was very young on an astronomical time scale. In 1939 Duncan had shown that the nebula was expanding at such a rate that it appeared to have originated in a point source only about 766 years earlier. But the most spectacular evidence of its youth was obtained from ancient Chinese and Japanese astronomical records, which described the appearance of a bright new star in the right part of the sky in the year AD 1054.

The extensive material in ancient and mediaeval Chinese records of comets and novae is described by Ho Peng-Yoke (1962). Records of a nova in the year 1054 were noted by Lundmark (1921), but the association with the Crab Nebula seems to be due to Duyvendak (1942). The history of the Sund Dynasty (Sung Shih, completed in 1345) contains this record: 'On a chi-chou day in the fifth month of the first year of the Chih-Ho reign-period a "guest star" appeared at the SE of Thien-Kuan (Taurus), measuring several inches. After more than a year it faded away'. The date is well corroborated in other independent records. The new star was visible in the day-time for several days, and remained as an object visible with the naked eye at night-time for nearly two years. There is no doubt that the ancient records describe a supernova explosion, and the near-coincidence of positions and dates makes the identification with the present-day Crab Nebula certain. The discrepancy between the actual birth date of AD 1054 and the date obtained by projecting back the presently measured velocities, which now converge at AD 1140 ± 10, is to be interpreted as a small but definite acceleration of the outward velocities, a fact of great significance in the question of the energy supply to the nebula.

Near the centre of expansion there are two stars, of 15th and 16th magnitude, which show prominently on good photographs of the nebula.

9.8 The Crab Nebula

Since there must be somewhere within the nebula a supply of energy to account for the continued emission of light, it was supposed that one of these two stars was the source of excitation. The one nearest to the centre of expansion was the south preceding star of the pair; it was also a star with a most unusual spectrum. Baade suggested that this might be the parent star for the whole nebula, but he was unable to account for the excitation through the familiar process of ultraviolet light emission. The spectrum showed no emission or absorption lines, which suggested that the star had a very high temperature. But at the same time there was no indication of the abnormally low colour index which would correspond to the strong ultraviolet emission of a hot star. We know now that this star is the Crab Pulsar, and that it feeds energy into the nebula not by light but through high-energy particles accelerated in a rotating magnetic field.

The Crab Nebula is contained within an ellipse 180×120 arcseconds across. The outer parts are filamentary, forming a network enclosing the more luminous central part. This is an amorphous mass concentrated towards the centre but extending over most of the major diameter and about two-thirds of the minor diameter. The light from these two components is totally different in character, so that in a colour photograph of the nebula the filaments show as predominantly red, while the centre is white or bluish-white. The red light from the filaments is line radiation, mainly Hα but including many other lines such as [N II], [O I], [O II], [O III], [S II], [Ne III], He I and He II. (The square brackets indicate forbidden transitions.) The relative abundance of the elements is close to the standard solar composition, except for a rather higher helium abundance. The ionisation is due to the ultraviolet light from the nebula.

The line radiation from the centre of the nebula originates in filaments on the front and back, i.e. the parts that approach and recede from the observer with maximum velocities. The Doppler shifts in these lines correspond to expansion velocities close to 1000 km s^{-1}. By combining this value for the expansion in the line of sight with the measured angular rate of expansion, and assuming that the expansion follows a simple elliptical form, the distance of the nebula may be obtained. This measurement and other estimates of the distance place the nebula in the range 1.5 to 2.5 kpc.

The white light from the central amorphous component has no spectral lines, and its origin was for a time a complete mystery. Although this component usually appears to be amorphous, under good seeing conditions it is found to be concentrated in fine filaments, like cotton wool. These fibrous concentrations run in organised directions, which are now known to be associated with a magnetic field within the nebula. The spectrum, and the high brightness, of this source of continuous radiation are incompatible with thermal radiation.

9.9 The continuum radiation from the Crab Nebula

In 1949 the radio astronomers J. G. Bolton, G. J. Stanley and O. B. Slee identified the Crab Nebula as a radio source. This was the first identification of a galactic radio source. The radio flux density was greater than in visible light, so that the continuum radiation could not be explained in the familiar terms of thermal radiation from ionised gas.

The explanation of this bright continuum radiation was provided by I. Shklovsky in 1953. High-energy electrons moving in a magnetic field follow curved paths; this curvature implies an acceleration, which leads to radiation. Previous analyses of energy loss in a synchrotron electron beam had already shown that this was an important effect, and it is generally called synchrotron radiation. It is also known as magnetic braking radiation, magnetobremsstrahlung. The main characteristics of synchrotron radiation are outlined in Chapter 14. The importance of this suggestion was that it provided the only known means for a very hot gas to radiate efficiently and over a wide range of wavelengths. Furthermore it led to the prediction that the radiation at any wavelength would be at least partly linearly polarised.

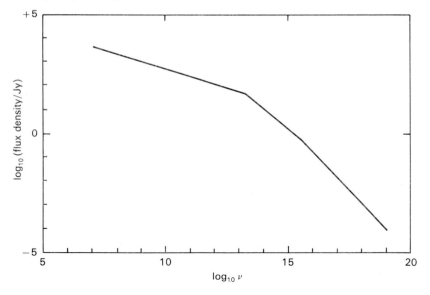

Fig. 9.4. The continuum spectrum of the Crab Nebula from radio to gamma-rays (Woltjer 1987).

9.10 The energy supply

Table 9.1. *Crab Nebula spectrum*

Frequency range (Hz)	Flux density (Janskys)
$10^7 < \nu < 10^{12}$	$F_\nu = 1040\, (\nu/10^9)^{-0.30}$ Jy
$2 \times 10^{13} < \nu < 3 \times 10^{15}$	$F_\nu = 1.82\, (\nu/10^{15})^{-0.85}$ Jy
$10^{16} < \nu < 10^{19}$	$F_\nu = 1.25 \times 10^{-3}\, (\nu/10^{18})^{-1.15}$ Jy

From Woltjer 1987.

Confirmation of the synchrotron proposal soon came from observations by two Soviet astronomers, Vashakidze (1954) and Dombrovsky (1954), who showed that there was indeed a large linear polarisation. A detailed investigation by Oort and Walraven (1956) was reported in a classic paper. Photographs in this paper show that the polarisation is so high that the appearance of the nebula varies dramatically according to the setting of a Polaroid filter on the telescope. The integrated light from the whole nebula is 9% polarised, while locally the polarisation may reach 60%. Oort and Walraven showed that the white radiation from the nebula must indeed be synchrotron radiation; their analysis showed that the magnetic field strength in the nebula must be about 10^{-3} gauss, and the electron energies must extend at least up to 10^{11} eV. The radiation mechanism was now understood, but the origin of the magnetic field and of the very high electron energy was to remain a mystery until the discovery of the Crab Pulsar.

Radio observations of the Crab Nebula now extend over the wavelength range from 30 m to 3 mm. The optical observations have been extended into the infrared, covering 500 nm to 5000 nm. X-ray and gamma-ray observations are now made from rockets and satellites, over the energy range 0.5 keV to 500 keV, corresponding to a wavelength range 3 nm to 3 pm. The known spectrum therefore covers a range of 10^{13}:1, or 43 octaves, spanning frequencies from 10^7 to 10^{20} Hz. A review by Woltjer (1987) shows that the spectrum is probably continuous, with the spectral indices and flux densities in the three main parts of the spectrum as shown in Table 9.1 and Fig. 9.4.

9.10 The energy supply

The analysis by Oort and Walraven of the optical and radio emission led to a fairly precise definition of the energy spectrum and actual numbers of electrons within the nebula, as well as the average value of the magnetic field. The total energy of fast particles in the nebula was found to be of the order of 10^{49} erg, most of this being concentrated in particles with energy of order 10^{11} eV. This energy is about one-thousandth of the total energy that would be released if a solar mass burned from hydrogen into helium, which is the maximum amount of nuclear energy that could reasonably be expected from a supernova explosion. It would be remarkable, but not inconceivable, for the energy of the explosion to be so well concentrated into high-energy particles. But a further problem was pointed out by Oort and Walraven. The electrons must be radiating so efficiently that their lifetimes are only of order 100 years rather than 1000 years, so that they could not have been accelerated in the original explosion.

The lifetime of an electron radiating synchrotron radiation with a maximum spectral density at frequency v(Hz) in a field B(gauss) is expressed as a half-life:

$$t_{1/2} = 10^{12} v^{-1/2} B^{-3/2} \text{ seconds.}$$

The lifetime could only be extended for optical radiation at a fixed frequency by assuming a smaller value of the field B, which would imply a larger total electron energy to produce the observed emission. The total energy would then reach or exceed the total available from the supernova explosion. This dilemma was made far worse by the observation of X-ray emission at energies in excess of 10 keV, where the synchrotron radiation must have come from electrons with energies of at least 10^{14} eV, which would have lifetimes of less than a year in any reasonable magnetic field.

An equally difficult problem is presented by the existence of the magnetic field itself, which contains the same order of magnitude of energy as the particles. Although there is no loss through radiation, it is impossible that this field is merely a remnant of a field that simply originated at the time of the supernova explosion. Any such field would have been reduced far below 10^{-3} gauss in an adiabatic expansion, transferring most of its energy into expansion energy of the nebula. There must be a means for the continued generation of a magnetic field throughout the nebula.

There was, therefore, even before the discovery of the Crab Pulsar, incontrovertible evidence that an energy source existed within the Crab Nebula that was providing both the high-energy particles and the magnetic field throughout the nebula. The location of this source was suspected to be at, or close to, Baade's star, both on account of its unusual spectrum and because of some remarkable activity in the nebula close to the star.

The activity close to Baade's star had been noted in 1921 by Lampland. Later observations, and especially some by Scargle and Harlan (1970),

9.10 The energy supply

confirmed his suggestion that some nebulous wisps about 10 arcseconds away from the star were moving and changing in brightness, sometimes even within periods of only a few months. If these were bulk movements of material, they must be moving at speeds approaching the speed of light. Between the wisps and the star there seemed to be a relatively empty space.

We can now be certain that the Crab Pulsar itself is supplying energy to the nebula, sufficient to provide for the magnetic field, the particle energy, and the accelerated expansion. The total energy required is about 10^{38} erg/s, which is close to the rate at which the pulsar is losing rotational energy.

10

Binary and millisecond pulsars

10.1 A separate population

The binary and the millisecond pulsars are in a different category from the general population. We have seen that the majority of pulsars are following a simple course of evolution, from a birth in a supernova, through a slowdown during which the magnetic field decays, to a death when the radiation ceases or becomes undetectable. The binary and millisecond pulsars constitute a separate and much older population. They result from an interaction with binary partners; most still have their companions, but others have lost them or are in the process of losing them, becoming solitary millisecond pulsars.

The interaction between the binary partners is a transfer of mass. This is a process which can be observed in the X-ray sources (Chapter 11). Here the mass transfer provides the thermal energy for the intense X-rays; from the point of view of pulsars the essential parameter is the transfer of angular momentum from the binary orbit to the pulsar. The result is a spin-up to a rotation speed of some hundreds of rotations per second.

The evolution of the binary and millisecond pulsars is to be considered separately from the majority of pulsars. The distinction is clearly seen in the P–$\dot{P}$ diagram (Fig. 10.1) where they all appear towards the lower left corner of the diagram. The characteristic ages $\frac{1}{2}P/\dot{P}$ of the millisecond pulsars are of order 10^9 years: this long lifetime is consistent with an origin in spin-up from old pulsars which the magnetic field has decayed. The lowest known value of $\dot{P}$ is in PSR 1953+29; here the values $P = 4$ ms, $\dot{P} = 3 \times 10^{-20}$ imply that this pulsar slows down by only one rotation in 16 months as compared with uniform rotation (i.e. with $\dot{P} = 0$).

10.2 Circular and elliptical orbits

In contrast to the common occurrence of binary systems among main sequence stars, the first hundred pulsars to be discovered were

10.2 Circular and elliptical orbits

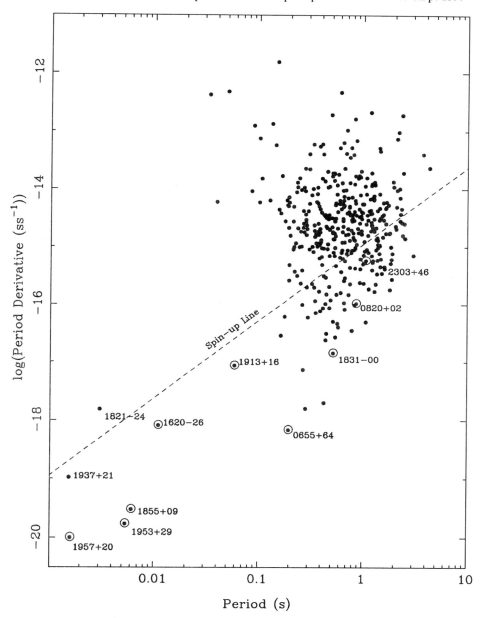

Fig. 10.1. The position of the millisecond and binary pulsars in the P–$\dot{P}$ diagram. The binaries are shown as open circles. The spin-up line is referred to on p. 139.

Table 10.1. *The binary pulsars*

PSR	P (ms)	$\log \dot{P}$	P_b (days)	e	Probable companion mass	Discovery
0655+64	196	−18.2	1.03	<0.00001	0.4–0.9	D1982
0820+02	865	−16.0	1232	0.0119	0.2–0.4	M1980
1620−26	11	−18.1	191.4	0.0253	0.3–0.4	L1988
1831−00	521	<−17.0	1.81	<0.005	0.06–0.13	D1986
1855+09	5.4	−18.8	12.3	0.00002	0.2–0.4	S1986
1913+16	59	−17.1	0.32	0.6171	1.4	See Ch. 5
1953+29	6.1	−19.5	117.35	0.0003	0.2–0.4	B1983
1957+20	1.6	—	0.38	<0.001	0.02–0.03	F1988
2303+46	1066	−15.4	12.34	0.658	1.2–2.5	S1985

References: Damashek *et al.* 1982; Manchester *et al.* 1980; Lyne *et al.* 1988; Dewey *et al.* 1986; Segelstein *et al.* 1986; Boriakoff *et al.* 1983; Fruchter *et al.* 1988; Stokes *et al.* 1985.

obviously solitary. This was easily demonstrated by the absence of periodic variation of the pulse periodicity; this is a very sensitive indicator of any orbital motion. The first pulsar to be found in a binary system, PSR 1913+16, has such a large orbital velocity and short orbital period of 7¾ hours that the rapid changes of pulse period due to changing Doppler shift made its detection and confirmation particularly difficult, and it is still possible that the commonly used search techniques (Chapter 3) discriminate against the rapidly changing periods of binaries in orbits with very short periods.

The statistics of ordinary binary star orbits show no preference for very low values of eccentricity. It is remarkable therefore that six of the first eight binary pulsar systems to be discovered have closely circular orbits (the exceptions being the highly eccentric system of PSR 1913+16, and PSR 2303+46). The main characteristics of all nine known up to mid-1988 are as shown in Table 10.1.

In none of the cases was it difficult to detect the binary nature of the pulsar, but the longest and the shortest periods were initially difficult to measure. For PSR 0820+02, the first detection (Manchester *et al.* 1980) was a short observation in which the changes of period were not apparent. Later observations, made to measure the rate of change of period $\dot{P}$, showed a negative value, suggesting a binary orbit, but the long period of nearly four years required an extended series of observations before the effects of the orbit could be separated from the intrinsic value of $\dot{P}$. The orbit is very nearly circular, with eccentricity $e = 0.012$. The combination

10.2 Circular and elliptical orbits

of a large orbit, in which two components are separated by a distance of the order of 1 AU, and a very low eccentricity, is an important indication of a past history of interaction.

The binary PSR 0655+64 must also be the result of a major interaction. Here the orbital period is only 24.7 hours (Fig. 10.2), and the eccentricity e is less than 10^{-5}. The closeness of the period to 24 hours led to some difficulty in the initial measurements of the orbit. It happened that the timing observations were made on the NRAO transit telescope at Greenbank, on which observations could be made for only a short period at about the same time each day. During the first few days of observations the behaviour was hard to understand, as the period appeared to be decreasing rapidly during each observation, but without a corresponding accumulated change. The orbit was only completely covered after more than a month of daily observation.

The orbit of PSR 0655+64 is known in more detail than is normally available from the Doppler variation of periodicity. In conventional optical observations, binary star systems are studied via spectroscopic Doppler shifts, which are closely analogous to the modulation of pulse period in the binary pulsars. It is a familiar result (see Section 10.5) that the inclination i of the orbit axis to the line of sight cannot be found from such measurements alone; instead the analysis gives a value of $a \sin i$, where a is the semi-major axis of the ellipse. A large orbit seen nearly face-on ($i=0$)

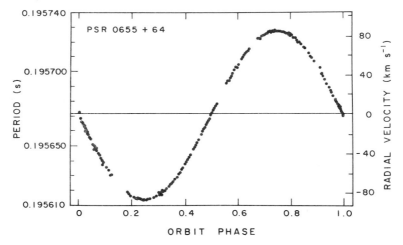

Fig. 10.2. Orbital velocity curve for PSR 0655+64. The orbit is very nearly circular, so that the observed pulsar period, reduced to the centre of the Solar System, various sinusoidally. The binary period is 24.7 hours (Damashek *et al.* 1982).

132 Binary and millisecond pulsars

cannot be distinguished from a smaller orbit seen edge on ($i=90°$), unless the transverse component of the orbital velocity can be measured. This has been achieved for PSR 0655+64 through measurement of radio scintillation rates, as discussed in Chapter 17. The variation of the transverse velocity through the orbit is directly related to the inclination; the measurements so far show that the inclination i is at least 60° (Lyne 1984).

As discussed in Chapter 5, the orbit of PSR 1913+16 is known in great detail through analysis of the relativistic effects in its highly elliptical orbit.

10.3 The globular cluster pulsars

The distinctive nature of the millisecond and binary pulsars first became apparent with the discovery of PSR 1937+21 (Backer *et al.* 1982). This solitary pulsar, with a period of only 1.6 milliseconds and a very low value of period derivative $\dot{P}$, has a characteristic age of 4×10^8 years, about 100 times greater than the typical age of normal pulsars. It was seen to be an old pulsar, which must have acquired its rapid rotation late in its life. The theory of accretion and spin-up in binaries was already current in connection with the X-ray binaries, and it was an obvious step to suggest that PSR 1937+21 should be classed with the three binary pulsars known at that

Fig. 10.3. The location of PSR 1821−24 in the globular closter M28. The radius of the circle is 20 arcseconds. The pulsar is shown by a cross in the expanded inset.

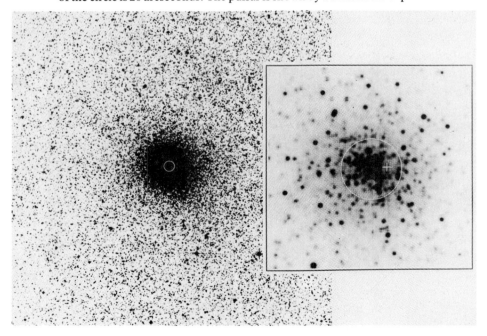

Table 10.2. *The globular cluster pulsars*

PSR	P (ms)	B_b	e	Location	Ref.
0021−72A	4.5	32 min	0.33	47 Tuc	A1988
0021−72B	6.1	7–95 days	?	47 Tuc	A1988
1620−26	11.1	191 days	0.025	M4	L1988b
1821−24	3.1	—	—	M28	L1988a
2127+11	110.7	?	?	M15	W1988

References: Ables *et al.* (1988); Lyne *et al.* (1988a); Lyne *et al.* (1988b); Wolszcan *et al.* (1988).

time, even though they had longer periods. All four might have been derived from the interacting binary systems which are observable as X-ray sources. It was also known that low mass X-ray binaries (LMXBs) were often located in globular clusters, which therefore became the targets of several intensive searches for millisecond pulsars.

Searches over large areas of the sky for millisecond pulsars have so far been rather unsuccessful, and many searches have been confined to special locations such as supernova remnants and globular clusters. A pulsar, PSR 1821−24, with a 3-millisecond period was eventually found in the globular cluster M28 (Fig. 10.3; Lyne *et al.* 1987). This cluster was found to contain a radio source with the steep spectrum and high polarisation characteristic of pulsars, but the source was comparatively weak, and it was only after an extensive search that the periodicity was discovered. The recording from which the discovery was made was a data set containing 5×10^8 samples, and the search for the periodicity took five hours on a large Cray XMP computer. A few months later another millisecond pulsar, PSR 1620−26, with a period of 11 milliseconds, was found in the globular cluster M4 (Lyne *et al.* 1988). This pulsar is in a binary system with a comparatively long period (191 days); the orbit is nearly circular. Table 10.2 lists the characteristics of the five pulsars found in globular clusters up to mid-1988: four of these are in binary systems, including two in the globular cluster 47 Tucanae.

These discoveries showed that LMXBs and millisecond pulsars are both common in globular clusters, giving strong support to the theory that they are associated in an evolutionary sequence.

10.4 The eclipsing millisecond binary

There is only a small chance that the orbit plane of a binary pulsar would be close enough to the line of sight for the pulsar to be occulted by its white dwarf companion, whose diameter of around 10^4 km is much less than the orbit diameter of, say, 10^7 km. The millisecond pulsar PSR 1957+20, discovered by Fruchter et al. in 1988, is occulted by its companion for one tenth of its binary orbit. The pulsar period P is 1.6 ms, and the orbital period P_b is 8 hours. The mass of the companion, presumably a white dwarf, is only about 0.02 $M_\odot$, but the duration of the occultation shows that the occulting disc is larger than the Sun. Immediately before and after the occultation there is a large increase in the dispersion measure of the pulsar, showing that the line of sight is passing through ionised gas. This explains the large size; the white dwarf is at the centre of an extended ionised atmosphere.

This ionised atmosphere can only be sustained by energy derived from the pulsar itself, probably through bombardment of the white dwarf by energetic charged particles. The white dwarf must be evaporating, and may only exist for another million years or so. Here we see in action a process which must eventually produce a solitary millisecond pulsar.

10.5 The masses of the binary pulsars and their companions

The general relativistic effects on the orbit of the binary PSR 1913+16 show that the masses of both components are 1.4±0.1 $M_\odot$. The only other neutron star masses which are deducible from observations are those of the massive X-ray binaries (Chapter 11), which are in the range 1.4 to 2 $M_\odot$. As we have seen in Chapter 2, theory allows a range of masses from 0.2 $M_\odot$ to 2 $M_\odot$, and it would be interesting if the masses were actually concentrated towards the top of this range. Can we expect further information from the other binary pulsars?

Classical dynamics gives only a combination of m_p, the pulsar mass, and m_c, the mass of its companion. Furthermore, the inclination i of the orbit is unobtainable from Doppler measurements alone, so that the combination usually available is the mass function

$$f(m_p, m_c, i) = \frac{4\pi^2}{G} \frac{a^3 \sin^3 i}{P_{\text{orb}}} = \frac{m_c^3 \sin^3 i}{(m_p + m_c)^2}. \qquad (10.1)$$

For example, the binary PSR 1953+29, with $P_{\text{orb}} = 120$ days, has a mass function of 0.00272 $M_\odot$. The possible combinations of mass are shown in Fig. 10.4, in which m_p is set at 0.5, 1.4 and 2.5 $M_\odot$. Evidently, m_c is unlikely to be less than 0.2 $M_\odot$: it cannot be greater than 0.5 $M_\odot$ unless $i \leq 20°$, but such a possibility cannot be excluded.

Similarly for PSR 0820+02 the probable range of mass of the companion lies between 0.2 $M_\odot$ and 0.4 $M_\odot$, although larger values are not excluded.

10.5 The masses of the binary pulsars

For the binary PSR 0655+64 we know the inclination i, and therefore we can assign unequivocal values to m_c for the whole possible range of m_p, as follows (Lyne 1984)

m_p	0.5	1.0	1.4	2.0
m_c	0.4	0.6	0.7	0.9

These masses for the companion of both these binaries, and for several others (Table 10.1), indicate that they are white dwarf stars, and therefore identifiable optically. Three have now been identified. Kulkarni (1986) found that the companion of PSR 0655+64 is a cool white dwarf, apparent magnitude $m_R = 21.8$, and that the companion of PSR 0820+02 is a hot white dwarf with $m_R = 23.2$. The difference between these two is important: the cool star must be very much older. Wright and Loh (1986) found that the companion of PSR 1855+09 is another cool white dwarf, also presumably an old star.

These identifications confirm our estimates of mass, and allow us to divide the binary pulsars between two groups: two with companions with a high mass, typical of a neutron star, and the others with low mass, typical of a white dwarf. Table 10.1 shows that the high mass binaries include the very eccentric orbits, and the low mass binaries include the longer orbital

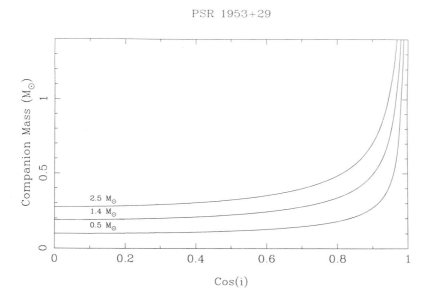

Fig. 10.4. The masses of the components of the binary PSR 1953+29. The curves show the mass m_c of the companion as a function of $\cos i$ for various masses m_p of the pulsar.

periods. This distinction is a vital clue to the evolution of the binary pulsar systems.

10.6 Evolution of binary systems

The X-ray binary systems, the binary pulsars and the individual pulsars, many of which may be presumed to be the remnants of former binary systems, together give a new understanding of a phenomenon, which appears to be common to the life of many stars: the interaction by mass transfer between members of a binary system. We first review the observational data.

The X-ray binary systems, which are the subject of Chapter 11, may be summarily described as follows. There are about 20 known such binary systems, with pulse periods from 0.15 to 853 s: these are the rotation periods of the neutron star. Many of the companions are visible stars, so that the masses are known. There are two groups, distinguished by the mass m_c of the companion. The more massive companions with $m_c > 10$ $M_\odot$ or thereabouts, are giant or super-giant stars: in these systems the companion has an extended envelope, which is gently spilling over on to the neutron star. The less massive systems, with m_c usually less than 2 $M_\odot$, are of later spectral type (A, F or G as compared with O or B): the neutron stars of these systems appear to be surrounded by an accretion disc that is fed by outflow from the companion.

The orbital periods of the more massive systems vary between 2.1 days (Cen X–3) and 35 days (1223–62) for the strong X-ray sources, with longer periods for some sporadic sources such as X Per (581 days). The low-mass sources have shorter periods, ranging from 41 min (1627–673) to 9.8 days (Cyg X–2).

There are many X-ray sources that are not positively identified as members of binary systems, but it is a reasonable assumption that most if not all have late-type stellar companions.

The pulsars themselves enter the scene as the nine binary systems already discussed, in the two groups of high mass, eccentric, short orbital period systems, and low mass, long orbital period, circular systems. The two single millisecond pulsars, with rotation periods 1½ and 3 ms, are closely related to the binaries: their rapid rotation must have been derived from a former binary interaction. The rest of the pulsars, which are single but which may have originated in binary systems, supply evidence through their measured velocities, which are notably widely distributed between low values (~ 10 or 20 km s^{-1}) and high values (~ 300 km s^{-1}). We must also take into account the life-time of the ordinary radio pulsars, which is around 10^7 years, and of the millisecond pulsars, which is of order 10^9 years.

10.7 Stellar evolution

The evolution of a solitary star follows a course which is mainly determined by one single parameter: its mass. The familiar main sequence of stars across the Hertzsprung-Russell Diagram is a sequence in which the smallest masses are represented by the cool faint stars, and the highest masses by the bright O and B stars. There are stars in formation that have not yet reached the main sequence, but most of the stars lying off the main sequence are in process of evolution after a long period of comparatively static existence on the main sequence. The age at which they turn to this more rapid evolution, and the rate of that evolution, both depend on their mass. A star like the Sun burns hydrogen quietly for about 10^{10} years before it exhausts its normal fuel supply and starts its evolution. Stars with lower mass therefore do not reach that stage within the lifetime of the Galaxy. In contrast, a star with mass 20 $M_\odot$ leaves the main sequence after only 10^6 years. These more massive stars then follow a spectacular course while they adjust to the burning of helium and more massive elements as their essential fuel.

Two aspects of the evolution are important in the context of binaries: these concern the inner core, and the outer envelope, both of which develop after the star moves off the main sequence. Paradoxically, the star both shrinks to produce a core hot enough to burn the new fuel and expands because of the large energy flow from it; in doing so, it becomes a red giant with a radius 10 or 100 times $R_\odot$ (the solar radius), containing a concentrated core which is destined to shrink to a white dwarf with a radius only $1/100$ $R_\odot$, or a neutron star with a radius only 2×10^{-5} $R_\odot$. The exact sequence of events depends, as before, on the total mass.

As we have seen in Section 9.2, a solar type star, i.e. with mass 1 $M_\odot$, becomes a white dwarf without a supernova explosion; when it collapses, the outer envelope is expelled gently to form a planetary nebula. Stars with mass greater than 2 $M_\odot$ may collapse explosively, expelling a large proportion of their mass. In a restricted range, no more than between 8 to 15 $M_\odot$, the remnant from the core collapse is expected to be a neutron star.

10.8 Mass transfer

A solitary star evolves on a track determined by its mass: the more massive it is, the faster it evolves. As we have seen in Chapter 9, the normal evolutionary sequence is upset by mass transfer in binary systems. At some point between the two stars of a binary system the gravitational pull of the two stars is equal and opposite. If the envelope of a red giant star expands reach this point, it is said to 'fill its Roche lobe', and its material can spill over and fall onto the other star. Since the red giant phase is unavoidable for massive stars, this process of mass transfer is a potent factor in the evolution of close binary systems. Even low-mass stars can fill their Roche lobe if the two stars are close enough: this is evidently the case for the low-mass X-ray systems. For the high-mass stars that occur in the massive X-ray binaries, the transfer may be the result of an outward stellar wind, which is effective even if the Roche lobe is not filled.

Mass transfer is the key to the history of the binary pulsars. It accounts both for the very high spin rate of the pulsar, since orbital angular momentum can be transferred to spin through accretion, and for the very low ellipticity of all but two of the orbits, since interaction usually tends to circularise any binary orbit. Mass transfer also complicates the evolution of both stars in the binary and opens up new possibilities for the birth of pulsars.

The majority of the millisecond pulsars are in binary systems. Do these binaries originate from the process of mass transfer? Can we account for the solitary millisecond pulsars in terms of a subsequent disruption?

If the binary radio pulsars are the product of the X-ray binaries, which are divided between systems with high-mass and low-mass companions, we may expect to find two corresponding categories of binary pulsar. The evolution of these systems has been traced by van den Heuvel and Taam (1984). In the high-mass systems the massive companion evolves and explodes as a supernova, much as it would if it were solitary. A large part of the total mass may be blown out of the binary system: if more than half is lost, the system will disrupt, leaving a solitary neutron star. If it does not disrupt, it will contain the neutron star remnants of two supernovae, probably in a very elliptical orbit as for the binary pulsar PSR 1913+16.

The more numerous low-mass systems (the LMXBs) evolve more gently. The companion is not sufficiently massive to make a supernova; instead it collapses only to a white dwarf. The mass transfer on to the neutron star is slower, and the mass and angular momentum of the whole system are conserved. The result of this slower and more orderly evolution is that the orbit becomes circular, which is a notable characteristic of several binary pulsar orbits. An example is PSR 1620−26, located in the globular cluster M4 (Section 10.3), where the diameter of the orbit has increased during the mass transfer. In other systems the orbit may shrink rather than expand; this occurs if the mass transfer is more rapid, resulting

10.8 Mass transfer

in a common envelope round the two stars. The envelope drags on their orbital velocity and brings them closer together. An example is PSR 0655+64.

In both the contracting and expanding systems the transfer of matter onto the pulsar can account for its spin-up to very short periods. As originally suggested by Eddington, there is an upper limit to the rate of accretion on to a star: the energy released by the accretion must not be so large that radiation pressure on the accreting matter overbalances the force of gravity. This suggests that the accretion rate is limited by the rate at which the neutron star can lose energy; in this case the limit of rotation speed is determined almost entirely by the dipole magnetic field (van den Heuvel 1987), giving

$$P = (1.9 \text{ ms}) B_g^{6/7} \qquad (10.2)$$

where B_g is the dipole surface field strength in units of 10^9 gauss. Since the dipole strength is related to the observables P and $\dot{P}$, we can plot a limiting line in the $P/\dot{P}$ diagram. This is the 'spin-up line' in Fig. 10.1; it forms a limit to the observed distribution.

The evolving companion stars usually become white dwarfs. Starting with a high mass, and losing a large part to the common envelope that is ultimately lost from the binary system, the companion becomes a 1 $M_\odot$ carbon/oxygen white dwarf, representing the degenerate core of a giant system with the outer envelope removed. Starting with a low mass, the degenerate core becomes a helium white dwarf with mass 0.5 $M_\odot$ or less.

The low values of dipole field deduced for the pulsars in most binary systems suggest that they are old enough for their original dipole fields to have decayed by factors of between 10 and 1000, i.e. that they are around 10^8–10^9 years old. This is a reasonable age for the contracting binaries, since their companions started as more massive stars, with a lifetime of that order. The expanding binaries, in contrast, started with low mass companions, and their evolution must have been so slow that the neutron star field should have decayed very much further. If this evolutionary scheme is correct, we must deduce that the decay of the field stops when the polar field has decayed to the order of 10^8 to 10^9 gauss. Another problem is to account for PSR 0820+02, which is in a large circular orbit with a white dwarf companion. According to the theory, this pulsar should be old; its magnetic field, however, is 3×10^{11} gauss instead of about 10^9 gauss as expected.

If it is indeed a young object, it may not have followed the evolutionary track from a supernova explosion; instead it may be the product of the accretion-induced collapse of a white dwarf, as described in Section 9.5.

A further possibility for the evolution of a close binary system is that the orbit may collapse as a consequence of gravitational radiation losses. Van den Heuvel and Bonsema (1984) pointed out that if the collapsing pair consisted of a neutron star and a white dwarf, the white dwarf will be disrupted and part of its mass will be accreted by the neutron star. The result will be a rapidly rotating solitary pulsar. This is evidently a possible origin of the solitary millisecond pulsars PSRs 1937+21 and 1821−24. The process is slow, but it does apply to all close binaries whatever their composition; in PSR 1913+16, which consists of two neutron stars, we actually observe the loss of orbital energy through gravitational radiation (Chapter 5). Whether it is the main origin of solitary millisecond pulsars, or whether the ablation effect observed in the eclipsing binary PSR 1957+20 occurs more frequently, is still unknown.

11

The X-ray binaries and bursters

In the same way that the pulsars belong almost completely to radio astronomy, and were a totally unexpected product of exploration of the radio sky, so the X-ray binaries belong to the new era of X-ray astronomy which began with the early rocket explorations of the 1950s. The first strong discrete X-ray source, Sco X–1, was totally unexpected and some years elapsed before the explanation was given by Shklovsky (1967). Confirmation of the nature of this and other X-ray sources in the Galaxy revealed by the UHURU satellite survey came when the source Cen X–3 was shown to be pulsating with period 4.8 seconds. Following the same arguments as in the interpretation of the pulsars, it soon became clear that the source must be a rapidly rotating neutron star.

Apart from the Crab Pulsar, whose X-ray pulses are closely similar to its radio pulses, the X-ray emission from these pulsating sources has an origin fundamentally different from that of the pulsars. The shape of the beam, as revealed by the 'light-curve' of the X-ray signal (Fig. 11.1), and the spectrum, both show that the source is thermal radiation from a local very hot region on the neutron star. It is astonishing that a small patch of hot gas, perhaps only 1 km across and 100 m thick, can produce enough thermal radiation to stand out by comparison with the background of hot stars: the reason is that the gas is very hot, at around 10^7 K, and optically thick, and that there is a very potent energy source to maintain such a concentration of energy.

All the pulsating X-ray sources are found to be in binary star systems; this is manifested by Doppler frequency shifts of the pulse period and the corresponding sinusoidal variations in arrival times such as those in Fig. 11.2. The orbital periods are typically several days, indicating that the binary systems are close enough for mass transfer to occur. In many cases the companion star can be seen optically, so that a complete resolution of the characteristics of the system including the masses of the separate components can be made. The companions are often massive early-type stars, whose evolution is expected to lead to expansion and overflow on to the surface of the neutron star. The energy supply is derived from the gravitational potential through which the overflowing matter falls.

No thermal radiation has yet been detected from an isolated neutron star, even though the sensitivity of EXOSAT was marginally sufficient to detect thermal radiation from young, nearby neutron stars, whose surface temperature might be around 10^6 K. The thermal energy of an isolated neutron star originates both in the gravitational collapse and in the continuing release of energy in the glitches. The surface temperature then depends on the conductivity of the interior and the emissivity of the surface; the magnetic field may affect the conductivity of the surface, so that the poles have a greater conductivity than the equator, although this does not give rise to an appreciable temperature difference over the surface (Hernquist 1985).

In this chapter we set out in more detail the observations and arguments that reveal the physical nature of the binary X-ray sources, and attempt to relate them to the binary pulsars discussed in the previous chapter.

11.1 Optical companions of X-ray binaries

Of approximately 100 optically detectable objects known in 1984 to be associated with compact X-ray sources, more than half are binaries. The others are supernova remnants, isolated hot white dwarfs, some non-

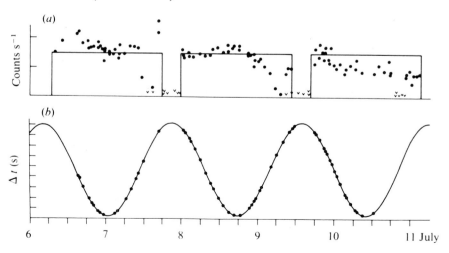

Fig. 11.1. The X-ray source Her X–1. (a) The cycle of X-ray intensity (the X-ray 'light-curve'), showing the periodic occultation; (b) sinusoidal variation of pulse arrival time (Giacconi 1974).

11.2 The massive X-ray binaries (MXBs)

degenerate stars, and some galactic clusters, the latter probably containing binary or more complex X-ray sources. We discuss only the binary systems: a comprehensive account of all optical identifications is given by Bradt and McClintock (1983). There are two fairly clear categories: those with massive early-type stars as companions, and those with low-mass ($M < M_\odot$), late-type stars as companions. Examples are listed in Table 11.1. More is known about the massive systems, which comprise 25 out of the 58 binary systems.

11.2 The massive X-ray binaries (MXBs)

In six of the massive systems the two masses can both be determined unequivocally, giving the only measurement of neutron star masses apart from the binary system PSR 1913+16 (Chapter 5).

The Doppler shifts of pulse period give values for the mass function (Chapter 8). Where the velocity of the companion can also be followed through the orbit by observation of optical absorption lines, the ratio of the peak-to-peak velocity ranges K gives the ratio of the masses M directly:

$$\frac{M_c}{M_n} = \frac{K_n}{K_c}$$

where the suffixes refer to the companion and the neutron star. For a circular orbit the total mass is given by the period P, apart from the

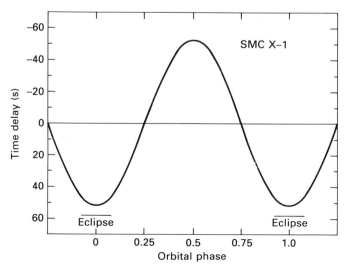

Fig. 11.2. Doppler shifts of pulse period through the binary cycle in SMC X–1, an X-ray source in the Small Magellanic Cloud (from Primini *et al.* 1977).

unknown inclination i of the orbit, since from Kepler's law:

$$G(M_c + M_n)\sin^3 i = (P/2\pi)(K_c + K_n)^3$$

where G is the gravitational constant.

Only the inclination i now needs to be determined; fortunately this is often available because the massive star, with its expanded envelope, eclipses the neutron star. Fig. 11.3 shows the effect for Cen X–3. A simple derivation of i assumes that the occulting star fills its critical potential lobe (the 'Roche lobe'), and the length of the eclipse then gives the inclination i. Generally, the existence of an eclipse allows a statement that $i > 45°$ approximately, so that errors in the precise geometry are not serious. An example is 0900–40, with period 8.96 days, inclination $i = 76° \pm 10$, in which $M_n = 1.9 \pm 0.3$ $M_\odot$ and $M_c = 24 \pm 2.5$ $M_\odot$ (Joss and Rappaport 1984). All such systems yield similar masses for the neutron star, and companion masses appropriate to massive O and B stars (Table 11.1).

11.3 The low-mass X-ray binaries (LMXBs)

The low-mass companions of the second group of X-ray binaries are more difficult to study, because little if any light can be observed directly from the star, and so there is no absorption-line spectrum from which to measure Doppler shifts. Light comes instead from gas in the accretion disc heated by X-rays, and, for some binaries such as Her X–1 and Cyg X–2 from the surface of the companion, again heated by the X-rays from the neutron star. Orbital periods have been obtained for some

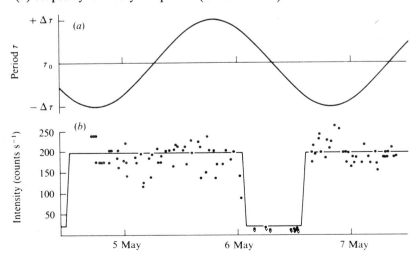

Fig. 11.3. The binary X-ray source Cen X–3. (a) Variation of pulse delay; (b) eclipse by its binary companion (Giacconi 1974).

11.3 The low-mass X-ray binaries (LMXBs)

Table 11.1. *Binary X-ray systems*

	Companion mass	Rotation period	Orbit period
Massive systems			
SMC X–1	17±4	0.7 s	3.9 days
Cen X–3	20±4	4.8 s	2.1 days
Vela X–1	24±2	283 s	9.0 days
4U 1853–52	20±8	529 s	3.7 days
LMC X–4	(19)	14 s	1.4 days
Low-mass systems			
Her X–1	2.3±0.3	1.2 s	1.7 days
Sco X–1	(1.2)	—	0.8 days

from variations in optical brightness, and some from Doppler shifts in X-ray pulse periods.

The orbital parameters cannot be obtained in the same detail as for the massive systems, but some progress can be made by assuming that the companion star fills its Roche lobe, and adopting a model relating the size of that lobe to the masses of the two stars. Unfortunately the model depends on the relation between the mass of the companion and its radius, and this relation depends on the type of the companion, which is usually unknown. Nevertheless, assuming the companion is on the main sequence, the model gives its mass; assuming further that the mass of the neutron star is 1.4 $M_\odot$, as in the massive binaries, we can obtain reasonable values for all the orbital parameters. Some typical values have been used in Fig. 11.4 to illustrate the geometry of massive and low-mass systems.

We should not assume that all companions of the low-mass binaries are of the same type. There are, however, several K-type main sequence stars clearly identified among them, including the companion of Cen X–3. The emission from this system varies between a low state, in which the K star can be identified, and a high state, in which the star is obscured by brighter radiation from the disc. This behaviour suggests that the other bright binaries are basically similar. Their distribution over the sky also supports the proposition that there is a concentration towards the centre of the Galaxy, following the distribution of the old stars of Population II. The low mass binaries are indeed often referred to as galactic bulge or late-type X-ray binaries. It seems likely that the X-ray sources in galactic clusters are of the same type.

Identification of many more of the low-mass binaries will be difficult. Many are found at low galactic latitude ($|b| < 2°$), where optical absorption

146 *The X-ray binaries and bursters*

is very heavy. Their spectral characteristics are very varied, although most have a large ultra-violet excess, by up to one magnitude over normal main-sequence stars. Their spectrum includes emission lines from He II ($\lambda 4686$) and N III ($\lambda 4640$), both generated by the X-rays from the neutron star. These line emissions, and the continuum, are notably variable from day to day.

11.4 Spin-up, accretion and inertia

In direct contrast to the pulsars, in which the periods are generally very stable apart from a slow smooth increase, the periods of the pulsating X-ray sources are decreasing rapidly and at an irregular rate. Figs. 11.5(*a*) and 11.5(*b*) show the periods of Her X–1 and Cen X–3 over two years. The timescale of the average increase, $P/\dot{P}$, is 3×10^5 yrs for Her X–1, and 3×10^3 yrs for Cen X–3. The torque responsible for this spin-up must therefore be more than three orders of magnitude greater than the magnetic torque on the pulsars, and in the opposite sense. This torque is provided by the accretion of material from the companion star via an

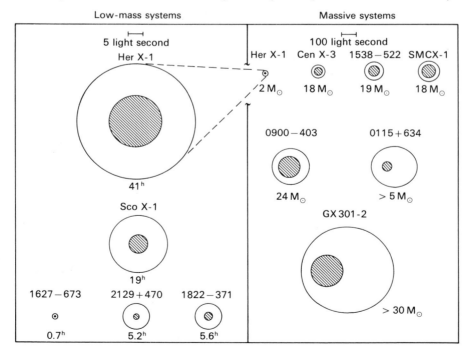

Fig. 11.4. The geometry of binary X-ray systems, showing the relative diameters and separations of low-mass and high-mass systems. The masses of the companions are indicated for the high-mass systems (Bradt & McClintock 1983).

11.4 Spin-up, accretion and inertia

accretion disc. The angular momentum transferred to the neutron star is derived from the orbital momentum of the binary system: the details of the transfer depend on the physics of the neutron star magnetosphere, to which we now turn.

The accreting material is heated as it condenses and is ionised into a plasma. As it falls towards the neutron star it encounters two important boundaries, the magnetospheric boundary, outside which the magnetic field of the neutron star is not felt, and the Alfvén surface, within which the plasma can only flow along the magnetic field lines. Between the two surfaces the flow pattern is determined by viscosity. From the point of view of spin-up torque, the flow becomes attached to the pulsar at the Alfvén surface. Outside the Alfvén surface the material is in Keplerian orbit round the neutron star, so that the accelerating torque on the star depends only on the rate of mass transfer and the radius of the Alfvén surface. The observed speed-up rates agree well with this theory, using the standard value of moment of inertia and a surface dipolar field of at least 3×10^{12} gauss.

The irregular nature of the spin-up seen in Fig. 11.5 must correspond to irregularities in the rate of mass transfer. There are also large fluctuations in the X-ray output, which may also be due to the same irregularities, although this is difficult to establish. An actual reversal of torque, causing

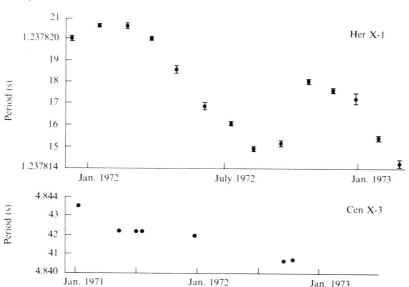

Fig. 11.5. Irregular decrease in the periods of Her X–1 and Cen X–3 (Giacconi 1974).

such a spin-down episode as that of 1972 Sept 10 in Cen X–3, requires a drastic change in the flow pattern, and probably a temporary ejection of matter from the neutron star.

11.5 X-ray light curves and a spectral line

The X-rays from the Crab Pulsar have the same basic pulse shape as the radio or optical pulses, and they can only be explained in terms of highly directional curvature radiation. The light curves of the binary X-ray sources are totally different; here the emitting region is a hot patch at the foot of the polar field lines down which the accretion flow is funnelled. This hot patch emits like a thermal radiator, and the X-ray 'light curve' seen as the neutron star rotates is, to first approximation, the variation in projected area of the hot spot. There are, however, many complications, as can be seen from Fig. 11.6. In particular, there are large differences between low and high energy X-rays; these differences are not consistent between different classes of X-ray sources, although they can be explained by more complex models of optical depth within the emitting region. In these models the polar cap may emit either a pencil beam or a fan beam, depending on the energy range.

There is also a very interesting observation of an X-ray spectral line near 55 keV in the spectrum of Her X–1. This has been interpreted by Trümper et al. (1978) as a cyclotron resonance in the emitting region. If this can be accepted despite some doubts about the precise conditions which would give such a sharp emission line, it gives a value for the cyclotron frequency in the emitting region, and hence a value of 6×10^{12} gauss for the polar

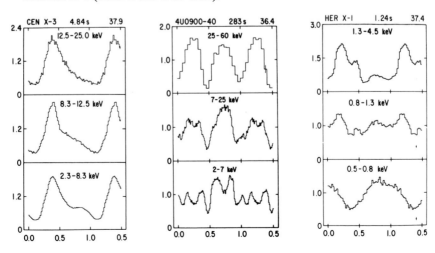

Fig. 11.6. X-ray light-curves at various wavelengths for Cen X–3, 4U 0900–40, and Her X–1 (from White et al. 1983).

11.6 The X-ray bursters

magnetic field. Confirmation might be obtained through measurements of X-ray polarisation.

Her X–1 is believed to be about 5×10^8 years old, and the existence of such a strong polar field is contrary to the widely accepted belief that neutron star magnetic fields decay with a time constant of 10^6 to 10^7 years (Chapter 8).

11.6 The X-ray bursters

X-rays from the galactic bulge sources are characterised by high intensity (total radiation $\sim 10^{34}$ erg sec^{-1}), soft spectra (T $\approx 3 \times 10^7$ K as compared with T $\approx 10^8$ K for the massive binaries), no pulsations, and no eclipses. In addition, a subset of the galactic bulge sources also show X-ray bursts, usually lasting some tens of seconds and recurring at intervals of some hours or days. These are the X-ray bursters. Their distribution in the sky is shown in Fig. 11.7, together with the known binary sources.

With some exceptions, the X-ray bursts are believed to be flashes of energy liberated in thermonuclear explosions in material accreted onto the surface of neutron stars. Although every astrophysicist readily accepts the concept of a controlled release of nuclear energy in the interior of normally evolving stars, as in a well-organised nuclear power station, it is a novel and startling concept to interpret the bursts as a sequence of nuclear explosions. The evidence, both observational and theoretical, is nevertheless very strong.

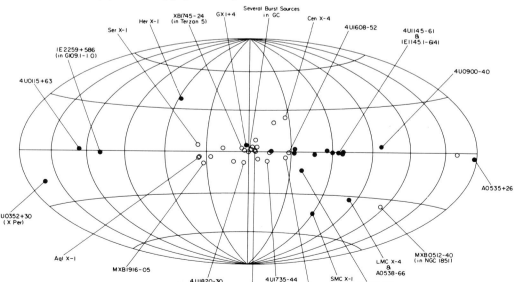

Fig. 11.7. Sky map of binary X-ray pulsars (filled circles) and burst sources (open circles) (Joss & Rappaport 1984).

The observed X-rays from a burst have a spectrum which fits well to a thermal source. The temperature of this source, however, falls during the burst, as seen by the progressive softening of the spectrum: typically 20-keV X-rays last for 20 seconds. The total energy, together with the temperature obtained from the spectrum, gives a measure of the area of the source, which is found to have a radius of about 7 km, corresponding to most or possibly all of the area of a neutron star. The temperature of this whole area is raised within a few seconds to about 3×10^7 K, cooling with a time constant of order 10 seconds. Some hours or days later the process is repeated.

The theory follows naturally from the known structure of neutron stars. The surface, shown in detail in Fig. 11.8 starts with a layer of order one metre thick, consisting of condensed hydrogen formed from the accreting plasma. As more hydrogen accretes on the surface, so must the hydrogen at the bottom of the layer fuse to form successively helium, carbon and iron, releasing energy as it progresses. There will therefore be a hydrogen burning region, followed by a helium layer also perhaps one metre thick, followed again by a helium burning layer. These thin burning layers, or shells, may be unstable if hydrogen is fed to the surface at a sufficient rate. The instability depends on the temperature coefficient of the fusion reaction.

The hydrogen-burning shell appears not to be the source of the instability, since hydrogen burning involves several intermediate stages of beta-decay from nuclei such as ^{13}N and ^{14}O which have time constants of 100 seconds or more. No such restraint exists in helium-burning, and it is the helium burning layer one or two metres below the surface that is the origin of the explosions.

The overall statistics of the energy release are simple. About 10^{21} grams of hydrogen accrete for each flash, corresponding to a million tonnes per square metre accreting over a time of about 10^4 seconds. The flash releases

Fig. 11.8. The surface structure of an accreting neutron star (after Lewin & Joss 1981).

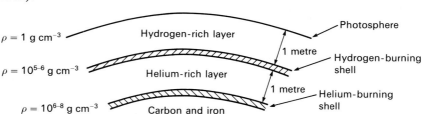

11.7 The Rapid Burster

up to 10^{39} ergs, which raises the temperature of the surface to 3×10^7 K in about 0.1 second. Radiation then cools the surface to below 10^7 K in about 10 seconds. Averaged over the cycle the mean power is 50 times that of the Sun: concentrated into a fraction of a second and into a small area, the event is seen as a thermonuclear explosion of truly awesome proportions.

11.7 The Rapid Burster

X-ray bursters were discovered independently by Grindlay *et al.* (1976), using the Netherlands Astronomical Satellite, and by Belian, Conner and Evans (1976) using records made several years previously by the Vela 5B satellite, which had been launched to monitor man-made nuclear explosions rather than the unexpected celestial bursts. A new type of bursting behaviour was discovered soon afterwards (Lewin *et al.* 1976) with the extraordinary property of a rapid succession on various time scales of the order of seconds to minutes (Fig. 11.9) from an X-ray source known as the Rapid Burster.

Fig. 11.9. The Rapid Burster X-ray source (from Joss & Rappaport 1984).

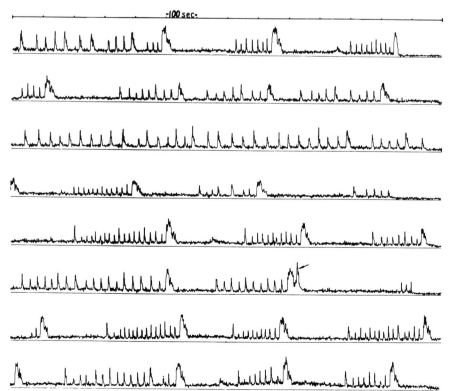

These rapid bursts represent an instability in the accretion rate rather than thermonuclear flashes. Both types of bursts can be seen from the Rapid Burster, as shown in Fig. 11.9. The rapid bursts, known as Type II, also represent very large releases of energy, but the source is gravitational rather than nuclear. About one tenth of the energy released per proton in the thermonuclear explosion of the slower Type I bursts is gravitational energy released in falling on to the surface.

11.8 Magnetic field decay

The most potent parameter which determines the differences in behaviour between the low- and high-mass systems is the strength of the dipolar magnetic field, which determines the pattern of the accretion flow on to the neutron star. A strong field channels the flow on to the poles: this flow is stable, and the nuclear fusion which occurs on the surface is, rather surprisingly, also stable at the high rate of flow. This is the pattern of the high-mass binaries, in which the X-rays originate in a small steady hot patch. A weaker field allows accretion over most or the whole of the neutron star, which is the pattern of the low-mass binaries.

This difference between the high magnetic fields of the massive binaries and the low magnetic field of the low mass binaries is presumably due to the different ages of the two classes. If we assume that neutron stars are always created with a strong magnetic field, which then decays with a time constant of order 3×10^6 years, the evolution of a companion low-mass star will be too slow for mass transfer to start before the field has effectively faded away.

Although many of the galactic bulge X-ray sources show neither the precisely periodic pulsations which indicate directly the presence of a neutron star, nor the clear spectrum of a late type star as companion, there is good reason for believing that all these galactic bulge sources are actually low-mass binary systems.

11.9 Quasi-periodic oscillations

The idea that mass transfer in binary systems, such as the LMXBs, would speed up the rotation of a neutron star component, received powerful support from the discovery of PSR 1953+29, which is a binary pulsar with period 6 milliseconds. The possibility that some of the LMXBs might contain such a rapidly rotating neutron star led to renewed efforts to detect short-period fluctuations in the flux of known X-ray sources. No such periodicity relating to the rotation of the neutron star could be found, but

11.9 Quasi-periodic oscillations

instead a totally different quasi-periodic fluctuation was found in several of the LMXBs.

This discovery is described in a review by Lewin and van Paradijs (1986). The quasi-periodic oscillations were found in intense X-ray sources including Sco X–1, Cyg X–2 and the Rapid Burster. The periodicity is variable; for Sco X–1 it varies between 6 Hz and 30 Hz, and in all sources this variation is strongly correlated with X-ray intensity. The periodicity is not precise; it is best described as a broad peak in a noise spectrum.

The accepted explanation of the quasi-periodic oscillations is that the period is determined by the Keplerian orbit period at the inner edge of the accretion disc. The diameter of the disc depends on the rate of accretion, which also affects the X-ray luminosity. High luminosity corresponds with a smaller diameter, and hence a shorter orbital period. Modulation of the X-ray luminosity must occur in the interaction of the rotation of the neutron star itself and the orbiting accretion disc, so that the observed frequency may be a beat between the two.

12

Integrated pulse profiles

Pulsars are like lighthouses, radiating a beam which sweeps across the observer at each rotation. The width and shape of the pulse of light from a lighthouse depend on the rotation speed and the angular width of the light beam. The precision of rotation speeds, the simplicity of the neutron star structure, and the isolation of the magnetosphere from all outside influence, combine to suggest that the radiated radio beams from all pulsars, as detected in the shapes of the received radio pulses, should all be similar. Although there is indeed a basic similarity, it is not at first apparent. The first complication was evident from the start of pulsar observations; there is an enormous variation between individual pulses received from any individual pulsar. This variation disappears when a sufficient number of pulses from the individual are superposed and added together to make an integrated profile. A sequence of only about one hundred pulses is needed to construct an integrated profile which is stable and characteristic of that pulsar. The second complication is that the integrated profiles differ greatly between one pulsar and another; this is partly due to the different ways in which the line of sight cuts the radiated beam, and partly due to irregularities within an individual pulsar beam.

While the majority of pulsars display very stable integrated profiles, a small number of pulsars switch between two different modes in which the integrated pulse profiles have different shapes. We describe this mode-switching phenomenon in Chapter 13, which is concerned with the more detailed behaviour of individual pulses.

The variety among the integrated pulse profiles constructed in this way can only be understood by integrating not only the intensity but also the polarisation parameters of the pulses. This understanding follows from the original observation by Radhakrishnan and Cooke (1969) that the plane of polarisation of the integrated profile of the Vela Pulsar swings smoothly through the pulse over an angle approaching 180°. Here again the individual pulses may vary, but the ensemble of polarisation parameters is stable and characteristic of the individual pulsar. Further, the pattern of polarisation through the integrated profiles follows a simple scheme, which can be applied to all pulsars, and which indicates clearly a universal model for the geometry of the emitting regions.

In this chapter we discuss the observations and describe what we have learned about the shape of the emission beam, how it varies with radio frequency, and how it depends on period and evolves with time. The model that emerges from our analysis is mainly due to Rankin (1983) and to Lyne and Manchester (1988).

12.1 The integrated profiles

The integrated pulse profiles are obtained by superposing a sequence of some hundreds of individual pulses. This is achieved by sampling the radio signal at small time intervals, and superposing or 'folding' the sequence of samples at the period of the pulsar. Since the pulses are often highly polarised, it is necessary to ensure that the total intensity is recorded; consequently two orthogonal modes of polarisation must be received, separately detected, and added. The polarisation characteristics themselves are of great importance; an integrated polarisation profile is built up in the same way as the intensity by adding the Stokes' Parameters of each sample. The signal-to-noise ratio in the integrated profile improves with larger receiver bandwidths and integration times; it is, however, necessary to restrict the bandwidth for pulsars with large dispersion measures, since there would otherwise be a smearing effect, which would spoil the time resolution. Observations can nevertheless be made with the wide bandwidths necessary for good signal-to-noise ratio by dividing the band into a sequence of sufficiently narrow bands and adding the separate recordings with the appropriate time delays, as described in Chapter 3.

The integrated profiles of some of the strongest pulsars, recorded at 408 MHz, are shown in Fig. 12.1. The profiles are smooth curves, generally with up to three components, more rarely with four or five. They generally occupy between 2% and 10% of the period. We may ask first whether this fraction is independent of the period, so that the profile length may be expressed at least approximately in terms of an angular width corresponding to a rotation angle of the pulsar. This is shown to be roughly the case in Fig. 12.2; here the width is taken between points where the intensity is 10% of the peak, so as to include all separate components of the profile. The straight line in Fig. 12.2 corresponds to an angular width of 9°. With few exceptions the widths lie between 3° and 30°.

156 *Integrated pulse profiles*

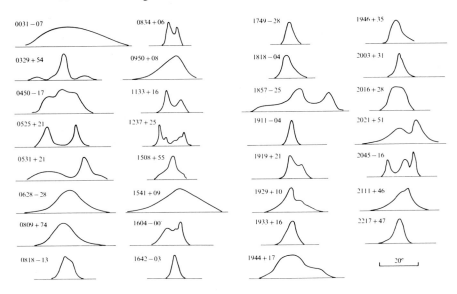

Fig. 12.1. Integrated profiles at 408 MHz of thirty-one pulsars, on a single scale of rotational longitude.

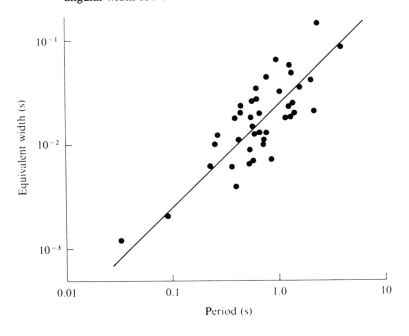

Fig. 12.2. Pulse widths plotted against period: the straight line corresponds to an angular width of 9°.

12.1 The integrated profiles

The separate components of the multiple profiles often appear to be discrete sources of emission; indeed the whole shape of the profiles can be interpreted as a sequence of discrete sources with a typical width of order one tenth that of the whole profile. The whole profile is, however, tied together by the polarisation. Six simple examples are shown in Fig. 12.3. Here the position angle of the plane polarised component swings in a characteristic S shape, covering a range up to 180°. Even in the multiple profiles this swing is continuous through the whole profile, suggesting that the emitting region is best delineated by the polarisation, which is determined by the configuration of the magnetic field. The separate components are then to be regarded as regions of excitation within a polar cap whose total extent and configuration are revealed by the polarisation.

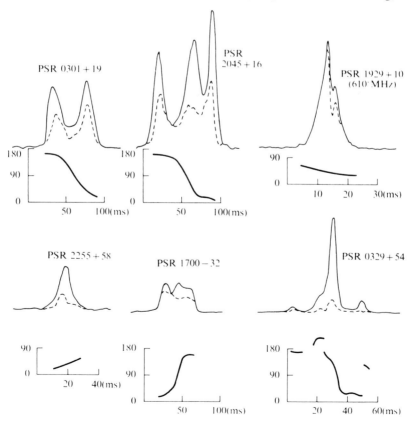

Fig. 12.3. Polarisation in integrated profiles. The broken line represents the linearly polarised component, and the graphs below the profiles show the position angle. The characteristic S-shape covers an angle up to, but not exceeding, 180°.

158 *Integrated pulse profiles*

Some components may be very weak or even completely missing, so that without the polarisation there could be no unequivocal interpretation in terms of a longitudinal distribution of emitting source in the pulsar magnetosphere.

Fig. 12.4. Integrated pulse profiles over a wide range of radio frequencies (Lyne & Manchester 1988).

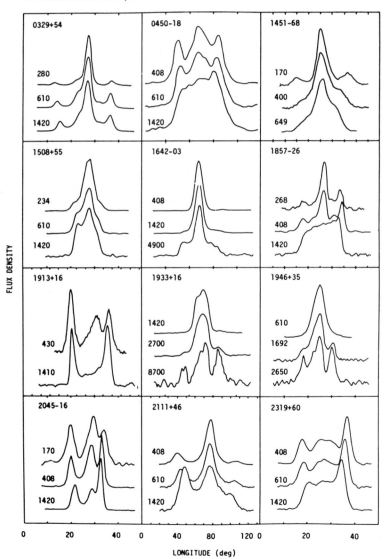

12.2 Radio frequency dependence

The shape and the width of all integrated profiles vary over the radio spectrum; some vary so much that profiles at, say 100 MHz and 10 GHz may be unrecognisably different, as in PSR 0329+54 and other examples in Fig. 12.4. The differences in shape are attributable to differences in spectral index between identifiable components: the differences in width are in some cases due to the addition or loss of outer components, but there is in addition a general expansion of the whole pattern at lower frequencies. Examples of this expansion are shown in Fig. 12.5. This expansion can often be fitted by a power law, in which the separation between identifiable components varies as $v^{-0.25}$ or thereabouts.

The overall radio spectra of the total intensity are invariably steep, although they often flatten or even turn over at lower frequencies (Fig. 12.6). If the flux density S varies with frequency according to the power law $S \propto v^{\alpha}$, the index α for the upper part of the spectrum lies in the range -2 to -4 (Sieber 1973). The indices of the separate components in an individual pulsar may differ from one another, varying over the same range; this accounts for the differences in profile shape. There is an organised difference in spectral index between the inner and outer components, so that for a triple profile such as PSR 2045−16 in Fig. 12.5 the outer components are prominent at high frequencies and are practically lost at low frequencies. Despite all these changes, the underlying swing of position angle

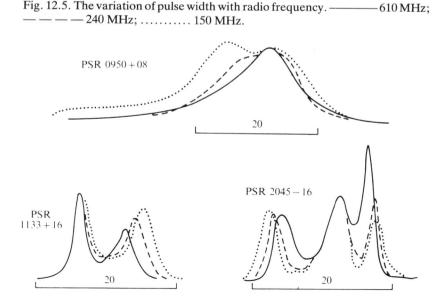

Fig. 12.5. The variation of pulse width with radio frequency. ———— 610 MHz; ———— 240 MHz; .... 150 MHz.

in the linear polarisation follows the same pattern at all frequencies, and provides the key to our understanding of the geometry of the emitting region.

12.3 Extended profiles and interpulses

Although the pulses from the majority of pulsars occupy only about 10° of rotation phase, there are some whose pulses extend more broadly. An example is PSR 0826−34, in Fig. 12.7(a). In some extreme examples, such as PSR 0950+08 in Fig. 12.7(b), a broad pulse profile may appear as two widely spaced components connected by weaker emission. In others, such as PSR 0823+26 and PSR 1702−19 (Fig. 12.8), there are two pulses which are distinct and separated by about 180°; both components behave like the single pulse of most pulsars. The weaker of the two is known as an interpulse. Naturally, the clear cases of interpulses are interpreted as emitted beams from the two magnetic poles of an orthogonal rotator, in which the dipole is perpendicular to the rotation axis and

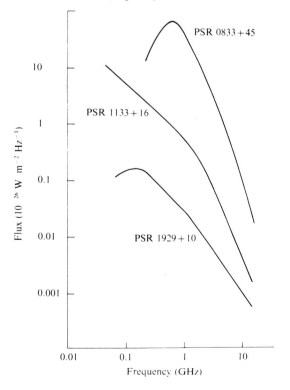

Fig. 12.6. Typical radio spectra of pulsars. The spectra are generally curved and often show a low-frequency cut-off.

12.3 Extended profiles and interpulses

Fig. 12.7. Wide pulse profiles of (a) PSR 0826−34 (b) 0950+08. The two peaks in these profiles were originally thought to originate in separate beams; they are in fact connected and represent the two edges of a single beam in a nearly aligned rotator.

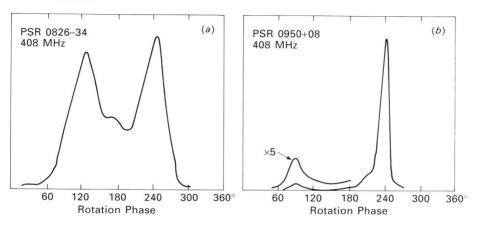

Fig. 12.8. Interpulses in PSR 0823+26 and PSR 1702−19. These represent the two distinct beams from the two poles of a nearly orthogonal rotator.

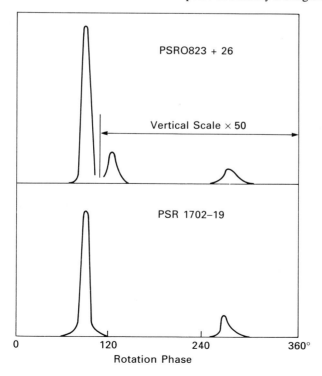

Table 12.1. *Pulsars with interpulses*

PSR	P (s)	$\dot{P}$ (10^{-15} s s^{-1})
0531+21	0.033	422.4
0823+26	0.531	1.72
0906−49	0.106	15.1
1055−52	0.197	5.83
1702−19	0.299	4.14
1736−29	0.322	7.9
1822−09	0.769	52.3
1855+09	0.0055	0.0001
1937+21	0.0016	0.0001

the observer is in the equatorial plane. This interpretation is confirmed by the polarisation characteristics, which show the typical S-shaped swing of position angle both for the main pulse and the interpulse.

Table 12.1 lists pulsars with interpulses, including only those whose polarisation characteristics indicate this orthogonal, bipolar configuration.

The values of P and $\dot{P}$ in the table show that the orthogonal bipolar configuration is found mainly in short period or young pulsars.

Other widely-spaced doubles, and some pulsars in which the emission fills most or all of the rotation period, have an entirely different geometry. Here the angle between the rotation and magnetic poles is small, and the observer sees the pulsar almost from the rotation pole. In these nearly aligned pulsars, the beam from only one pole is observed, but it is visible for a large part of the period. If the outer parts of the cone are brightest, a wide spaced double may be observed, as for PSR 0950+08.

12.4 The overall angular width

Even if the whole cone of emission could be seen, we would expect to observe large differences between the widths of integrated profiles of pulsars in which there are different angles of inclination between the spin and magnetic polar axes; similarly there should be differences between those observed at various 'impact parameters', i.e. those in which the observer's line of sight approaches more or less closely to the centre of the cone of emission. These factors have been taken into account by Lyne and Manchester (1988) by using the polarisation characteristics to determine the geometrical configuration for each of a large sample of pulsars.

Using this configuration, they obtain a corrected pulse width for each

12.4 The overall angular width

pulsar, i.e. the pulse width corresponding to a diametral cut across the radiated beam. The full pulse widths at 400 MHz, as indicated by the polarisation swing, are plotted in Fig. 12.9 as a function of period P. This plot does not take account of the inclination between the rotation and magnetic axes; if this departs from 90° the angular width of the radiated beam will be exaggerated in this plot.

The lower limit to the widths follows a line

$$\text{apparent beamwidth} = 13° P^{-1/3}$$

where P is the period in seconds.

We consider this to be the actual relation between radiated beamwidth and period for orthogonal or nearly orthogonal rotators. The wide spread of observed widths above this line corresponds to non-orthogonal rotators, in which the rotation and magnetic axes are more or less aligned. Lyne and Manchester deduce from the distribution of points in this plot that there is a tendency for pulsars with longer periods to be more aligned; this is presumably the result of an alignment process that is operating through most of a pulsar's lifetime.

Pulsars are only observable if the beam crosses the observer's line of sight. The total number of pulsars is therefore greater than the population deducible from the pulsar surveys (Chapter 8) by a 'beaming factor'; the

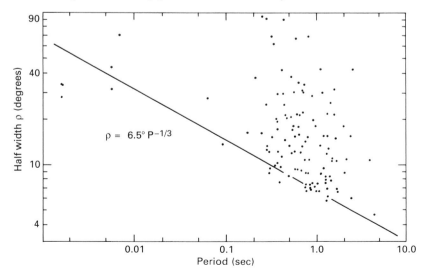

Fig. 12.9. The full pulse widths of the integrated profiles, corrected by reference to the polarisation swing. The lower limit corresponds to a beam width of $13° P^{-1/3}$, where the period P is in seconds. Pulsars above this line are non-orthogonal rotators (Lyne & Manchester 1988).

164 *Integrated pulse profiles*

beam widths and alignments found in this chapter suggest that this factor varies from about two or three for the short period pulsars to about four or five for the long period pulsars.

12.5 Circular polarisation

Towards the centre of many profiles, and especially those which cut close to the centre of the emitting cone, there is often a high degree of circular polarisation. Examples are shown in Fig. 12.10. The degree of circular polarisation often reaches 10%; it reaches 50% at the centre of the profile of PSR 1702−19. Frequently it reverses hand, as seen in PSR

Fig. 12.10. Circular polarisation (broken line, showing Stokes parameter V), seen near the centres of the integrated profiles of PSRs 1237+25, 1508+55, and 1702−19.

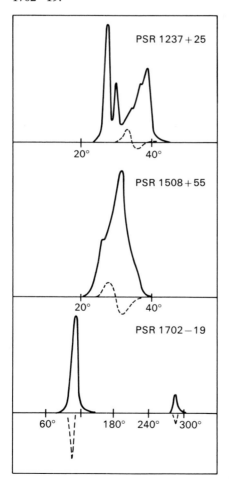

12.7 An empirical model of polar cap emission

1508+55; usually there is only one such reversal in the whole profile. As with most polarisation characteristics, this circularly polarised component appears to be almost invariant with frequency.

12.6 Orthogonal polarisation

As we will see in the next chapter, the polarisation of individual pulses may be very different from that of the ensemble average, i.e. the integrated profile. A particular manifestation of this is the occurrence of linearly polarised pulses whose plane of polarisation is at right angles to the average. In some profiles this orthogonal mode is dominant for part of the profile, and instead of the smooth S-shaped swing of position angle there is a discontinuous step in the curve. An example is shown in Fig. 12.11. This complication in interpreting integrated profiles is in some cases made worse by a variation across the radio spectrum of the proportion of pulses in the orthogonal mode; this can give rise to spectral differences in the polarisation characteristics, providing exceptions to the general rule that the polarisation is invariant with frequency.

12.7 An empirical model of polar cap emission

A consistent model, incorporating the descriptions in the preceding paragraphs, has been presented by Lyne and Manchester (1988), based on work by Rankin (1983). The essential elements of this model are:

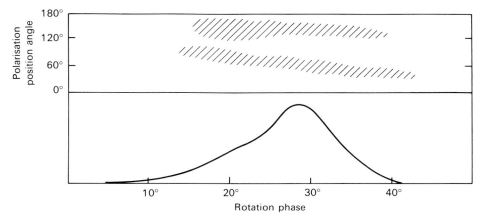

Fig. 12.11. Orthogonal linear polarisation. The smooth polarisation curve of the integrated profile is, in some pulsars, interrupted by a 90° step in position angle. The superposed plots from many individual pulses from PSR 0950+08 show two regions separated by 90° in which the position angle may lie (Backer & Rankin 1986).

(i) A radio beam is emitted from each magnetic pole, approximately radially, with an angular width of order 10°.

(ii) The dipole axis may be aligned at any angle with the rotation axis, and the observer's line of sight may be at any angle to the rotation axis. There is a tendency for the magnetic axis to align with the rotation axis, on a time scale of order 5×10^6 years, i.e. about the same time scale as that of the decay of the dipole field strength.

(iii) The beam is conical, with an overall width which increases with decreasing radio frequency. Particularly at high frequencies, the emission is often concentrated on the outside of the cone; a cut across the beam may then give a double or a single profile depending on the way the observer's line of sight cuts across the polar cap. There are also components of emission from within the cone; these may fill in more or less the double profiles,

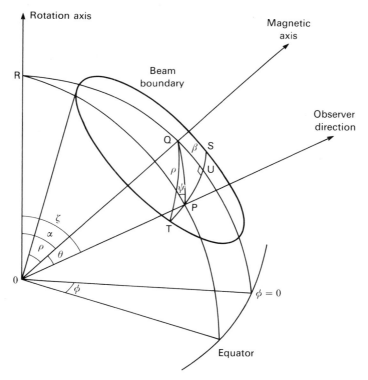

Fig. 12.12. The geometry of pulsar beams. The observer sees radiation from the point P, which moves across the arc ST as the pulsar rotates. The zero of longitude φ is defined as the meridian through the magnetic axis, and position angles ψ are measured with respect to the projected direction of the magnetic axis. The conical emission beam has an opening semi-angle ϱ. The swing of polarisation position angle depends on the impact parameter β.

12.7 An empirical model of polar cap emission

sometimes resulting in profiles with three, four, or five components. The shape of a profile depends on the 'impact parameter', as shown in Fig. 12.12, and on the relative strengths of the various components.

(iv) The spectral index varies regularly from the centre to the outside of the cone, so that the centre region or 'core' is relatively stronger at low frequencies. This difference in spectral index is most noticeable in pulsars with short periods.

(v) The cone of emission is patchy, so that only parts of it may be observable. A cut across the beam may therefore show only part of the typical pattern; for example it may give a double profile in which the first half is the outer cone and the second is the central core, the expected later part of the outer cone being missing. There is no clearly distinguishable 'core'; emitting regions may occur at any location within the cone.

(vi) The swing of linear polarisation across the profile always conforms to a simple pattern which is determined by the configuration of the rotation axis, the magnetic axis and the observer's line of sight. This pattern may, however, be obscured by an orthogonal mode of polarisation occurring in various parts of the profile.

(vii) Circular polarisation is observed when the line of sight is close to the magnetic pole, i.e. when the field lines have the lowest curvature. The reversal of hand suggests that a source that is confined to a small tube of field lines is observed from either side of the plane containing those field lines, so that the curvature of the field lines is seen to reverse as the line of sight crosses the plane. We note that this is characteristic of curvature radiation, and we explore the consequences in Chapter 16.

(viii) The angular width of the radiation cone increases at lower radio frequencies. In Chapter 14 we shall interpret this in terms of an emitting region located within an expanding cone over the magnetic pole, so that the lower frequencies are emitted in regions at greater radial distances from the pole.

13

Individual pulses

The well-organised and easily reproduced pulse profiles obtained by integrating some hundreds of individual pulses conceal a rich diversity of behaviour among the individual pulses. The integrated profiles are made up of very varied individual pulses, each of which may have more than one component: it is the statistical distribution of these components over a range of longitude, combined with their characteristic width and the probability distribution of their intensities that determine the repeatable shape of the integrated profiles (Fig. 13.1).

The components of an individual pulse are often identifiable as characteristic 'sub-pulses', with a typical width in longitude of 1° to 3° (as compared with the typical width of 10° for the integrated profiles). These sub-pulses may occur apparently at random longitudes within the 'window' defined by the integrated profile, or they may show a preference for certain longitudes at which the integrated profile shows a peak, or again they may 'drift' across the window, appearing at a longitude which changes slowly from pulse to pulse.

The sub-pulses are regarded as basic components of the pulse profile; an individual sub-pulse then represents the radiation from an isolated location within the distribution of locations covered by the integrated profile.

There is also structure on a much shorter time scale, known as the *'microstructure'*. This appears to be a modulation of sub-pulse radiation rather than a distinct component of radiation, although it does take the form of well-separated very sharp and intense 'micropulses'. Groups of micropulses, spaced approximately periodically, may occur within a single pulse.

All components, whether they are subpulses or micropulses, usually show a very high degree of polarisation. The polarisation characteristics of subpulses are organised in a simple manner, but the changes through a

Individual pulses

single subpulse, and the variations from one subpulse to the next are often sufficient to dilute the polarisation in the integrated profile.

In this chapter we describe these phenomena in the light of the geometric description of the source of pulsar radio emission which we set out in the previous chapter: i.e. a pattern of emission from a cone above a magnetic pole. At a given radio frequency the emission is from a definite height, which is greater for lower frequencies. The emission is not uniform over the cone; the outer ring is often a more powerful emitter, so that a cut across the cone can give a double pulse. The inner and outer components often have different spectral indices, so that the inner components become more prominent at lower frequencies. The emissivity is also patchy, so that

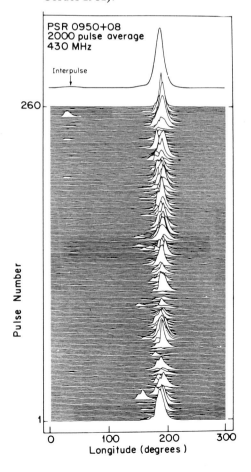

Fig. 13.1. A sequence of pulses from PSR 0950+08 with the integrated profile obtained by adding together the sequence of individual pulses (Hankins & Cordes 1981).

a cut across the polar cone may give a series of apparently discrete components.

13.1 Single pulse intensities and pulse nulling

We note first that the quoted mean intensity of a pulsar conceals an interesting distribution of total pulse power among the individual pulses (Fig. 13.2; Smith 1973). Many pulsars, such as PSR 1642−03 show a distribution of total pulse power which resembles a normal distribution about a mean value. Others, such as PSR 0950+08, show an asymmetric distribution with high probability at low values and a long tail towards high values of pulse power. The Crab Pulsar (PSR 0531+21) provides the extreme example of an asymmetric distribution, with very occasional 'Giant Pulses' (Chapter 7). PSR 0834+06 shows a bimodal distribution: there is a finite probability of zero power, and a separate distribution of values about a mean. The zeroes occur in groups of pulses; they represent a switching off to a level well below one per cent of the mean pulse power. The phenomenon is known as *'nulling'*.

The length of the nulls, and the interval between their occurrences, vary randomly about characteristic values. For some pulsars, nulls two or three pulses long may occur at intervals of order one hundred pulses (Backer 1970), while for others the null state may last for many minutes and occupy half the total time. The switch between the two states is very rapid. An extreme example of long nulls is found in PSR 0826−34, whose discovery

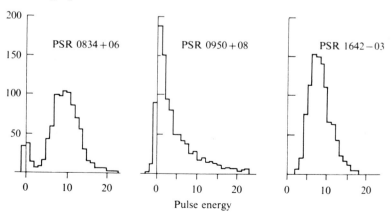

Fig. 13.2. The distribution of pulse power among the individual pulses from three pulsars, recorded at 408 MHz. PSR 0834+06 shows a small peak at zero intensity (the missing pulses). PSR 0950+08 shows a maximum at zero, with a monotonic fall. PSR 1642−03 has no missing pulses, and shows a smooth distribution about a single peak.

13.1 Single pulse intensities and pulse nulling

was very difficult to confirm because the pulsar proved to be in the null state for 70% of the time, with nulls extending for periods of over seven hours (Durdin *et al.* 1979). Short nulls may, of course, be missed in observations of weak pulsars which can only be detected after integration of many pulsars.

The occurrence of nulling appears to be a characteristic of the older pulsars. Most of the nulling pulsars are found in the P–$\dot{P}$ diagram (Fig. 13.3) close to the 'death line' (Ritchings 1976); pulsars approaching this line from the lefthand side are nearing the end of their active lives. A natural interpretation would be that pulsar emission ceases abruptly at this stage by the same process as is involved in nulling; possibly the cessation

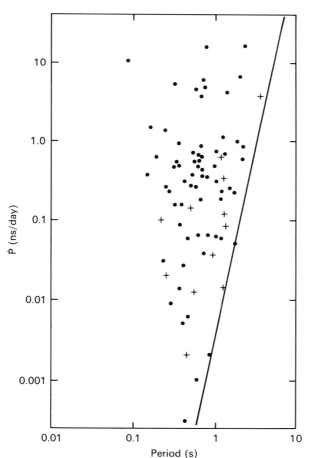

Fig. 13.3. The position of pulsars showing nulling or drifting in a logarithmic plot of period P against its derivative $\dot{P}$. These pulsars occur at low values of $\dot{P}$ and close to the cut-off line where radio emission ceases.

172 *Individual pulses*

takes place through stages of nulling in which the proportion of time spent in the nulled state becomes progressively larger. It is, however clear that the total radiated power decreases with age, whether or not nulling occurs; this decrease is probably directly related to the decay of the dipole magnetic field (see Chapter 8).

13.2 Mode changing

We have already noted in Chapter 12 the mode changing behaviour of many pulsars, in which the integrated profile switches between two different forms. Examples are shown in Fig. 13.4. A mode change may resemble a transition to a null state, since it may involve the cessation of radiation from one or more identifiable components of the integrated profile; we may therefore regard a null as a mode change in which all components have disappeared simultaneously. A mode change, however, seems to involve a redistribution of the excitation over several components, which usually have different radio spectra. Changes can also be observed in the behaviour of the sub-pulses, and particularly in their

Fig. 13.4. Changes in the integrated profiles due to mode changes in the pulsars PSR 1237+25 and PSR 0329+54 (Bartel *et al.* 1982).

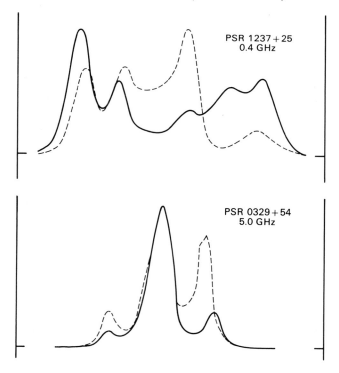

13.3 Sub-pulses

Table 13.1. *Sub-pulse widths and full integrated pulse widths*

PSR	$W°_{sub}$	$W°_{int}$	PSR	$W°_{sub}$	$W°_{int}$
0031−07	5.5	36	1642−03	2.7	7
0138+59	2.0	28	1706−16	1.2	12
0301+19	2.6	19	1749−28	2.9	8
0320+39	1.1	9	1818−04	4.6	11
0329±54	1.9	16	1821+05	2.7	20
0355+54	2.7	30	1831−03	4.6	34
0450−18	2.9	40	1857−26	4.1	42
0525+21	1.7	20	1900+01	6.4	13
0540+23	2.6	28	1907+10	6.3	14
0628−28	4.1	38	1911−04	3.1	6
0740−28	9.7	19	1919+21	1.6	11
0809+74	3.8	26	1920+21	2.9	13
0818−13	1.5	12	1929+10	2.0	24
0820+02	1.9	17	1933+16	1.9	12
0823+26	2.4	10	1944+17	6.5	37
0826−34	4.4	190	1946+35	6.4	24
0834+06	1.1	9	2002+31	1.9	5
0919+06	2.4	20	2016+28	2.4	14
0943+10	2.5	21	2020+28	2.9	18
0950+08	5.2	31	2021+51	2.4	22
1112+50	1.1	9	2045−16	1.1	17
1133+16	0.8	12	2111+46	9.1	11
1237+25	0.8	15	2154+40	3.3	26
1508+55	3.0	7	2217+47	3.1	11
1540−06	1.1	10	2303+30	1.3	8
1541+09	13.7	24	2310+42	1.4	16
1604−00	2.7	17	2319+60	2.4	25

Notes: Sub-pulse widths are between half-power points; integrated pulse widths are between 10% power points. Both are in degrees of rotation. The sub-pulse data were compiled by M. Ashworth at Jodrell Bank.

rate of drifting. For instance, the pulsar PSR 0031−07 exhibits three distinct modes, in which the sub-pulses drift at different rates, as well as a null; these different modes and the null occur in a sequence, showing that both types of switching phenomena are closely related.

13.3 Sub-pulses

The individual pulses of most pulsars are composed of one or more sub-pulses, each with a smooth, approximately Gaussian shape, with a characteristic width of about 2° of rotational longitude. This width is almost independent of radio frequency.

Table 13.1 shows the full width to half intensity of typical sub-pulses,

174 *Individual pulses*

together with the full width of the integrated profile (quoted here between 10% intensity points so as to avoid ambiguities in some of the more complex profiles). For some pulsars, such as PSR 1929+10, the two widths are nearly the same; individual pulses from this pulsar occur remarkably regularly at nearly the same longitude and with nearly the same shape. For others, such as PSR 0950+08 (see Fig. 13.1), the sub-pulses are much narrower than the integrated profile. In PSR 1133+16, which has a double-peaked profile, they occur preferentially at or near the longitudes of the peaks; the widths of these peaks are largely determined by the widths of the sub-pulses.

We interpret the sub-pulses as basic beams of radiation from discrete sources, and their longitude distribution within the pulse window as a physical distribution in longitude at the pulsar.

13.4 Drifting and modulation

In many of the long-period pulsars, such as PSR 0809+74 and PSR 0031−07 (Fig. 13.5), successive pulses contain sub-pulses which appear at

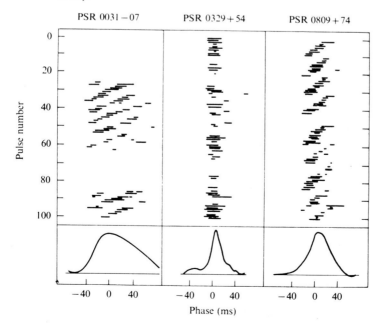

Fig. 13.5. Drifting and nulling. Each horizontal line is centred on the expected arrival time, with time increasing downwards and to the right. The positions of each sub-pulse are shown. PSR 0031−07 and PSR 0809+74 are typical drifters. PSR 0031−7 shows large nulls, missing about twenty pulses (Taylor & Huguenin 1971).

13.4 Drifting and modulation

progressively changing longitudes. This marching of the sub-pulses across the window of the integrated profile is shown in the idealised diagram of Fig. 13.6. The normal pulse periodicity is labelled P_1, and P_2 is the spacing between sub-pulses within a single pulse. The drift at rate D ($= P_2/P_3$) brings successive sub-pulses to the same longitude at intervals P_3. The drift periodicity P_3 is usually expressed in terms of P_1.

The cycle of repetition at interval P_3 may be observed as a modulation of the integrated pulse power (Fig. 13.7). This modulation is often more obvious if a restricted range of longitude is analysed. Fig. 13.8 shows the result of a Fourier analysis, at five discrete longitude intervals, of a long train of pulses from PSR 1237+25. Here the characteristic periodicity $P_3 = 3P_1$ shows a modulation frequency 0.35 cycles per period P_1. The drift is, however, only seen in the outer parts of the pulse window, indicating that the phenomenon is located in the outer cone of emitting regions.

There are several examples of pulsars in which P_3 is close to $2P_1$, with the result that pulses are alternately strong and weak. A Fourier analysis of the total pulse power of PSR 0943+10, for example, shows a dominant

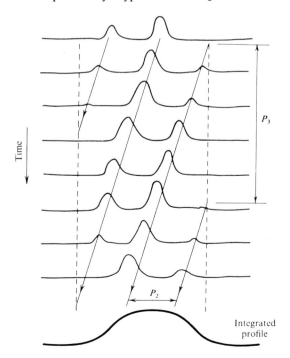

Fig. 13.6. An idealised pattern of drifting sub-pulses. Successive pulses appear at the fundamental period P_1. The pattern repeats at interval P_3. Sub-pulses are separated by a typical interval P_2.

176 *Individual pulses*

modulation at 0.473 cycles per P_1 (Fig. 13.7). In such cases it is difficult to distinguish this frequency from the alias frequency 0.527 cycles per P_1; the two possibilities correspond to drifting in different directions at slightly different rates.

Periodic modulation may be observed in the outer components of some pulsars without any apparent drifting. The best-known example is PSR 1237+25, shown in Fig. 13.8. This may be understood in the light of the polar cone model of the previous chapter. If the sub-pulses correspond to localised emitting regions on the outer cone, these regions are apparently drifting round the circular zone. For all but an exactly diametral cut across the beam, the subpulses are then drifting in longitude at a rate which

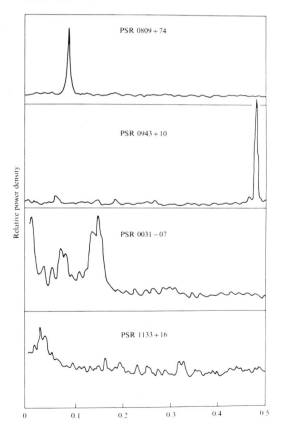

Fig. 13.7. Periodic intensity modulation due to frequency drifting. The figure shows a Fourier analysis of the intensities of a long train of pulses from four pulsars (Taylor & Huguenin 1971). The period P_3 is well defined in PSR 0809+74 and PSR 0943+10. Drifting is more complex in PSR 0031−07, and irregular in PSR 1133+16.

13.5 Drift rates 177

depends on the distance of the cut from the pole. For pulsars like PSR 1237+25, where the cut is almost exactly along a diameter of the beam, the emitting regions move in and out of the observed zone without drifting in longitude, giving modulation without drift.

13.5 Drift rates

Table 13.2 presents the drift rate D for pulsars in which drifting is consistently observed, in units of degrees of longitude per rotation. The table also shows the data for P_1, P_2 and P_3. The overall accuracy for D is about 10% for most of these pulsars. Some pulsars display more than one drift rate: in these cases the component spacing P_2 does not change, and the change in P_3 is inversely proportional to the drift rate. Two pulsars, PSR 0031−07 and PSR 2319+60, each show three distinct values of drift rate corresponding to three different modes. The three rates for PSR 0031−07 are almost harmonically related, suggesting the name 'The Harmonic Pulsar'; this may be accidental, but it is remarkable that the same ratios of drift rate are found, within the rather wide limits of observational error, in PSR 2319+60. In both pulsars the succession of the modes tend to follow a definite pattern including a null, as seen for PSR 0031−07 in Fig. 13.5.

The drift rate for several pulsars varies significantly through the range of longitudes covered by the integrated profile. In Table 13.2, two distinct

Fig. 13.8. Periodic modulation of intensity in a pulsar with no apparent simple drifting. In a series of pulses from PSR 1237+25 the outer components of the pulse vary with a cycle of 0.36 pulse periods (Taylor & Huguenin 1971). The inner components show no fluctuations.

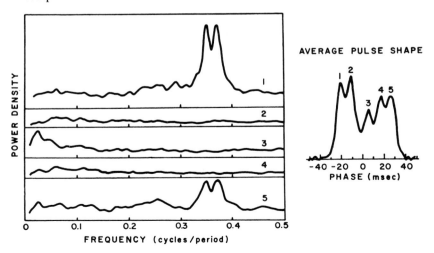

Table 13.2. *Drift parameters*

PSR	P_1 (s)	P_2 (ms)	P_3 periods	D (deg/P_1)	Ref.
0031−07	0.94	21	12.5	−1.7 ±0.2	WF
			6.8	−3.2 ±0.6	
			4.0	−5.3 ±1.1	
0148−06	1.46	32	14	−0.57	
			7	−0.30	BHMM
0301+19	1.39	24	6.4	−0.96	SS
0320+39	3.03	22	8.5	+0.31 ±0.2	MA
0525+21	3.75	27	4	−0.7 ±0.2	MA
0809+74	1.29	47	11.1	−1.2 ±0.2	URWE
0818−13	1.24	14.5	4.4	−3.0 ±0.1	MA
0820+02	0.87	16	4.9	+1.2 ±0.4	MA
0826−34	1.84	148	irreg	irreg	BMHML
0834+06	1.27	3.5	2.16	+0.25 ±0.16	MA
0943+10	1.10	25	2.11	+4.0 ±0.1	SO
1237+25	1.38	41(Cpt1)	2.8	3.8 ±1.0	BSW
		85(Cpt2)	2.8	7.3 ±1.5	
1540−06	0.71	4.6	3.07	0.8 ±0.15	MA
2303+30	1.58	22	2.0	−2.4 ±0.1	MA
2310+42	0.35	3.9	2.05	+1.8 ±0.3 (or −2.1 ±0.3)	MA
2319+60	2.26	38	8	0.8 ±0.2	WF
			4	1.3 ±0.6	
			3	2.0 ±0.7	

References:
BHMM = Biggs, Hamilton *et al.* (1985)
BMHML = Biggs, McCulloch *et al.* (1985)
MA = Ashworth, unpublished data from Jodrell Bank
WF = Wright & Fowler (1981)
SS = Schonhardt & Sieber (1973)
URWE = Unwin *et al.* (1978)
SO = Sieber & Oster (1975)
BSW = Bartel *et al.* (1980)

rates are quoted for two of the inner components of PSR 1237+25; this is an extreme example, and it is more usual to find a variation of order 20% in drift rate across the profile. Drifting is most commonly seen in the outer components of the profile, which correspond in the polar cone model to the outer cone of emitting regions. The drift rate is often appreciably higher at lower radio frequencies, where the emission is from higher up in the diverging cone; the periodicity P_3 is, however, the same, as would be expected if the emitting regions are connected along magnetic field lines. The drift rate in the inner cone, if it appears at all, is often lower than in the outer, and it may appear only as a slow fluctuation of intensity.

The direction of drift is indicated by the sign in the Table, where + means a drift from the leading to the trailing edge of the profile. Early observations by Ritchings and Lyne (1975) suggested a distinction between pulsars with high values of period derivatives $\dot{P}$, which show positive drifting, and pulsars with low values of $\dot{P}$, which show negative drifting. This distinction is now less clear; if it is correct, it may represent a difference between two configurations of the line of sight to the pulsar, in which the line of sight either cuts between the rotation pole and the magnetic pole or passes below the magnetic pole, as in Fig. 12.12. Such a difference might be related to $\dot{P}$ if the axes tend to become aligned during the lifetime of pulsars.

13.6 Drifting after nulling

Following our interpretation of the sub-pulses as discrete sources of emission, whose drift represents an organised movement round the polar cone, it is interesting to observe sub-pulse drifting immediately before and after a null. Cole (1970) showed that the emission restarted at a longitude 'remembered' from the point where the null started; Lyne and Ashworth (1983) observed this effect in more detail for PSR 0809+74, in which nulls are observed which last for between one and fourteen periods. In Fig. 13.9 the phase of the drift after a null would appear as a horizontal line if the drift were unaffected by the null. The plot is a compendium of measurements for 159 nulls of various lengths up to 14 missing pulses. The phase of the drifting is clearly related before and after a null; the relation depends, however, on the length of the null.

The pattern of drift before and after a null is interpreted by Lyne and Ashworth as follows. A small change in drift rate can be seen in the last ten or so pulses before the null. Immediately after the null starts the drift stops, and then recovers exponentially to its former value. The time constant of this recovery depends on the length of the null; it is about 10 seconds for short nulls and 20 seconds for long nulls. The emission restarts when the drift rate has partly recovered. For short nulls the drift does not completely halt.

Continuity of drift phase across an extended null implies the continued existence of an identifiable source, even when its emission is temporarily

180 *Individual pulses*

suppressed or absent. An attractive model for such a source is presented by Filippenko and Radhakrishnan (1982), who suggest that the spark discharge across the gap at the magnetic pole is confined to discrete flux tubes, each of which has a continuous flow of energetic particles. The emission from a flux tube depends on the energy distribution and the bunching of electrons or positrons in the flow; the flux tube and the continuous stream may therefore remain in place even if the radiation has stopped. The radiation starts again when the flow again becomes bunched.

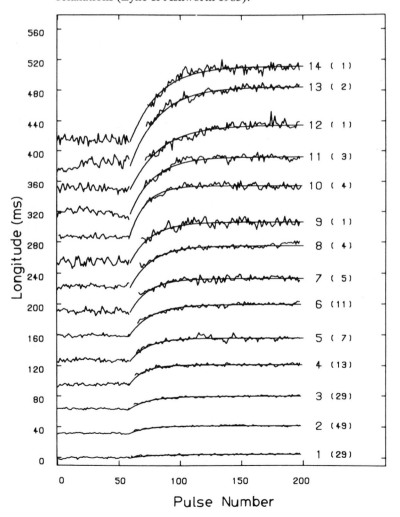

Fig. 13.9. The phase of the drifting sub-pulses at nulls of various lengths in PSR 0809+74. The null length is shown at the right of each track, followed in parenthesis by the number of tracks in each average. The curves are exponential relaxations (Lyne & Ashworth 1983).

13.8 Microstructure

Ruderman and Sutherland (1975) showed that a localised discharge across the polar gap would tend to move sideways; this drift would depend on the component of electric field **E** along the direction of the magnetic field **B** in the gap, giving a small **E**×**B** drift relative to the surface of the star. The change in emission at the time of the null, and the change in drift rate, would then both be due to a change in the discharge across the gap.

13.7 The polarisation of sub-pulses

Sub-pulses are generally more highly polarised than integrated profiles; the difference is accounted for by the variability of polarisation of successive sub-pulses and by variations within individual sub-pulses. Recordings of the polarisation of individual sub-pulses by Manchester, Taylor and Huguenin (1975) demonstrate these effects clearly. Fig. 13.10(a) shows a highly polarised pulsar, PSR 1929+10. Here the individual pulses are fully elliptically polarised, with little change through the duration of the sub-pulse. Fig. 13.10(b) shows a similar situation in PSR 0031−07, in which there is an organised drifting. Here the polarisation varies across the window of the integrated profile, as expected, but there are also some additional variations within the sub-pulses. These variations are often accounted for by switches between orthogonal modes (see for example Fig. 12.11).

Taylor *et al.* (1971) found that there is an organised sweep of position angle within individual pulses in addition to the sweep expected from the polarisation of the integrated profile. It seems that the total swing within a single sub-pulse is typically less than about 30°, in this and in other pulsars.

Finally, Fig. 13.10(d) shows PSR 1919+21, which demonstrates the confusion that arises from the superposition of several different sub-pulses in a single pulse. It is a reasonable assumption that the individual sub-pulses which contribute to such a pulse are highly polarised in a simple manner, and that a number of different sources are observed simultaneously.

13.8 Microstructure

At the highest possible time resolution, and more usually at low observing frequencies, the individual pulses of many pulsars are seen to contain structure on a considerably shorter time scale than the sub-pulses. This is referred to as the microstructure. Detailed observations are only possible for the most intense pulsars; even for these there is some difficulty in resolving the shortest pulse components, or 'micropulses', which may be only a few microseconds wide.

182 *Individual pulses*

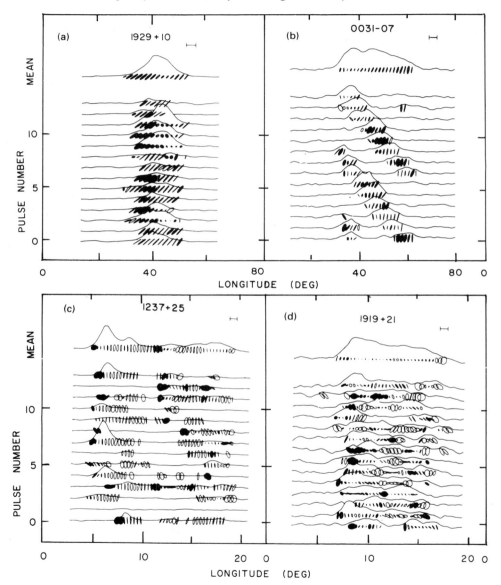

Fig. 13.10. The polarisation of individual sub-pulses from four pulsars, compared with their integrated profiles: (*a*) PSR 1929+10, (*b*) PSR 0031−07, (*c*) PSR 1237+25, (*d*) PSR 1919+21. The major axes of the polarisation ellipses represent the fractional polarisation and the position angle; left hand is represented by filled ellipses (Manchester, Taylor & Huguenin 1975).

13.8 Microstructure

Simultaneous observations at widely separated frequencies show that micropulses have a wide bandwidth. Rickett, Hankins and Cordes (1975), for example, showed that micropulses observed at 111 MHz and 318 MHz from PSR 0950+08 and from PSR 1133+16 occasionally contained structure shorter than 10 microseconds, which was correlated at the two frequencies. The polarisation of micropulses is generally similar to that of the underlying sub-pulse.

Micropulses are often observed to occur in short quasi-periodic groups, with a periodicity of order half to one millisecond. It is very significant that this periodicity is found to be independent of the observing frequency, as may be seen in the example of Fig. 13.11, showing a single pulse of PSR 0950+08 at 430 MHz and 1406 MHz (Boriakoff *et al.* 1981). Following the polar cone model, in which emitted radio frequencies decrease outwards along magnetic field lines, we can immediately interpret the microstructure as a variable excitation progressing outwards along a magnetic flux tube. We have already noted that sub-pulse drifting may be interpreted as the sideways drift of a discharge in such a flux tube; we now add to this model by proposing that the discharge is intermittent.

Fig. 13.11. A periodic train of micropulses from PSR 0950+08, recorded simultaneously at 430 MHz and 1406 MHz (Boriakoff *et al.* 1981).

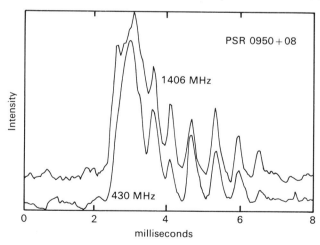

14

Geometry of the emitting regions

Although it was soon established that the pulsars were not pulsating but rotating, and that the radiation was emitted in a narrow lighthouse beam, it took many more years to discover the nature of the beaming process. No other celestial source, apart from the hydroxyl and water masers, produces such a narrow beam. No known radiation mechanism could produce a narrow beam over the whole of the electromagnetic spectrum, as observed in the Crab Pulsar. Two theories were explored extensively: these became known as the *polar cap* and the *relativistic beaming* theories. The first placed the source of the emission immediately above a magnetic pole; the other placed it far out in the magnetosphere, close to the velocity of light cylinder. In both theories the location and direction of emission were determined by the dipolar magnetic field.

The evidence for the polar cap model came from the radio observations of beamwidth and polarisation, while the evidence for the origin in the outer magnetosphere came from the high energy radiation observed from the young pulsars, Crab and Vela. It proved to be impossible to reconcile the two theories by suggesting that all radiation from the young pulsars came from a single location, and it is now accepted that there are two different origins. These are related to two different locations in the magnetosphere where charged particles can be accelerated to high energies. We consider first the radiation with high photon energy which originates in an outer magnetosphere gap.

14.1 The outer magnetosphere gap

The beaming mechanism that is responsible for an identical beam pattern over more than ten decades of the Crab Pulsar spectrum must be determined by an essentially simple geometrical constraint. The radiating region is confined to the outer magnetosphere gap, whose position is determined by the pattern of magnetic field lines immediately outside the

14.1 The outer magnetosphere gap

region of closed field lines (Fig. 2.2). The beam direction, and the timing of pulses received as the star rotates, are also affected by the relativistic velocity of the outer magnetosphere and by the travel time of radiation within the magnetosphere. The fact that pulse timing and pulse shapes are practically identical over the whole range of emitted wavelengths and photon energies indicates that there is one single pattern of emitting regions within the outer magnetosphere gap.

As we have seen in Chapter 2, the magnetosphere, in broad terms, consists of an extensive plasma envelope, much of it charge-separated, which is swept round by the magnetic field pattern rotating with the neutron star. Acceleration of particles can only occur in regions of the magnetosphere that are depleted of charge. Where there is total depletion, the induced electric field may be very high, reaching 10^{12} V cm^{-1} for the Crab Pulsar. The outer magnetosphere gap is a depleted region extending over a surface immediately inside the closed field lines. In the young pulsars this gap is traversed by gamma-rays generated lower in the magnetosphere. Within the gap these gamma-rays create pairs of electrons and positrons; this pair-creation occurs when the gamma-rays encounter either a transverse component of magnetic field or a considerable density of lower-energy radiation, both of which are present in the gap. The newly created particles can then be accelerated along the field lines to very high energies, radiating curvature radiation in a beam directed along the field lines.

The geometry of the emitting region therefore follows very closely the geometry of those field lines which delineate the edge of the polar cap. Smith (1986) analysed the simple case of the orthogonal rotator, in which the magnetic dipole is perpendicular to the rotation axis. The analysis has to take into account both the effect of rotation on the shape of the magnetic field lines and the relativistic aberration of the radiated ray as seen by a stationary observer on the equatorial plane.

The observed pulse arrival time is expressed as a phase ϕ given by

$$2\pi\varphi = (\theta - 90 + \alpha) - x \sin\alpha \tag{14.1}$$

where θ is the rotation phase of the dipole, α is the aberrated ray angle, and x is the radial distance of the source as a fraction of the velocity of light radius, as in Fig. 14.1.

The angle χ between the radius vector and the field line is given by

$$2\tan\chi = (1+x^2)\tan\theta + x \tag{14.2}$$

and the aberrated ray angle α is given by

$$\tan\alpha = \frac{(1+x^2)^{1/2} \cos\chi}{x + \sin\chi.} \tag{14.3}$$

The plot of arrival phase in Fig. 14.1 shows the arrival times for emission originating close to the limiting field lines as a function of x, the normalised radial distance. The zero of pulse phase is the expected arrival time for a 'polar cap' source, i.e. a source at a magnetic pole and at $x=0$. Only one of the two possible curves (from the two magnetic poles) is drawn.

Fig. 14.1 also shows the observed pulse profiles for the Crab and Vela Pulsars, including the radio precursor for the Crab. If only one pole were involved, the double pulse must be explained in this model by postulating that both the leading and the trailing edges of the polar cone radiate nearly equally, and that the radial distance x is in the region 0.8 to 0.95. A more realistic model involving both poles requires that only the leading edges radiate; in this case the simple orthogonal model cannot apply, on account of the unsymmetrical spacing between the pulses.

14.2 Non-orthogonal configurations

The interval of 0.4 period between the two high-energy pulse components was explained by Cheng, Ho, and Ruderman (1986) in terms of the light travel time across a large part of the magnetosphere. A source radiating both inwards and outwards, as might be expected from particles

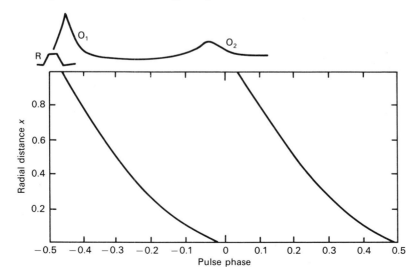

Fig. 14.1. Pulse arrival times for optical sources O_1, O_2 close to the limiting field lines of the polar cap and at radial distance which is a fraction x of the radial distance to the velocity of light cylinder. The radio pulse R originates close to the pulsar surface at the magnetic pole.

14.3 Polarisation of the high-energy source

with opposite charges accelerated in an outer gap, would be seen twice in each rotation; the geometry giving the correct time interval was easily constructed in this model. As already discussed, the difficulty in this model is that one of the two pulses would have to traverse a large part of the magnetosphere, where the gamma-ray pulse would be lost by pair creation.

A different explanation of the unsymmetrical spacing between the two pulse components is shown in Fig. 14.2. This shows a simple version of the geometry, in which the two sources of the high energy pulses are associated with the two magnetic poles but located close to the velocity of light cylinder. The non-orthogonal arrangement of the dipole, and placing the observer away from the equatorial plane, combine to give a light travel time which differs appreciably for the two sources. We return to this model after discussing measurements of the optical polarisation of the pulses from the Crab Pulsar.

14.3 Polarisation of the high-energy source

As described in Chapter 7, the optical emission from the Crab Pulsar is highly polarised. Unfortunately, this provides the only infor-

Fig. 14.2. A geometrical scheme for two sources located close to the velocity of light cylinder and associated with the two magnetic poles. Each source follows a circular track and is observed at the tangent points S_1, S_2. The delay between the two pulses depends on the inclination of the magnetic axis to the rotation axis, and on the position of the observer.

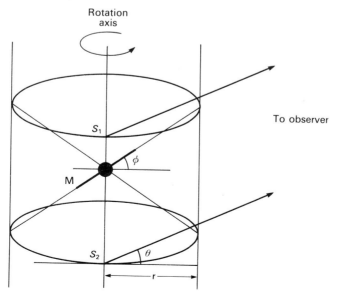

mation on the polarisation of the high-energy source; in the Vela Pulsar the optical radiation is too weak, while the techniques for X-ray and other polarisation measurements are not yet available in satellite telescopes. Fortunately, however, the optical measurements of the Crab (Smith *et al.* 1988) reveal a simple situation, which varies little if at all across the optical band; we may reasonably surmise that the same would be found over the whole of the high-energy region. The radiation is highly linearly polarised through most of the pulse period, and the position angle changes very little except near the two pulse peaks, where it swings smoothly through a range of about 90°. The same curve (Fig. 14.3) is followed for both components, indicating that the geometry of the two sources is identical. This favours the interpretation of Fig. 14.2 rather than the less symmetrical configurations.

The linear polarisation of curvature radiation is aligned in the plane of curvature of the magnetic field. Originating at the polar cap, this is in the form of an expanding cone; the high-energy radiation is generated at the wide-open trumpet-like mouth of the cone close to the velocity of light cylinder; furthermore the source is concentrated at the edge of the cone, in the outer magnetosphere gaps. The polarisation is determined by the apparent direction of the magnetic field in these outer regions. The

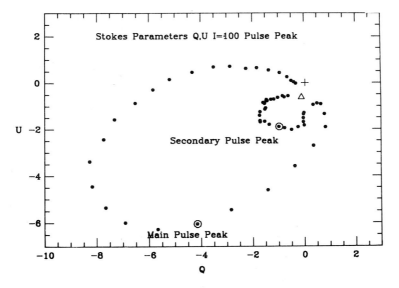

Fig. 14.3. Polarised optical radiation from the Crab Pulsar. The Stokes parameters Q, U are plotted at uniform intervals of pulsar rotation. The main pulse and 'interpulse' follow the same pattern of polarisation, but with different intensities.

geometry is complicated by the effects of aberration, since the source is moving with high relativistic velocity. Qualitative agreement with theory can be obtained over a range of source configurations; it is however essential that both the magnetic dipole axis and the observer are aligned at angles between 20° and 40° to the equatorial plane of the rotating pulsar.

14.4 The polar cap and the source of radio emission

As for the high-energy regime, the shape in time of the radio pulse profiles represents a spatial distribution of emitting regions on the pulsar, primarily in longitude. The polar cap model locates the emitting regions in the diverging pattern of field lines over a magnetic pole. In contrast to the high-energy regime, the source is fairly close to the star and is little affected by aberration; it also fills the polar cone rather than being concentrated at the leading edge. The radiating charged particles are accelerated in a vacuum gap over the magnetic pole rather than further out, in the outer magnetosphere gap.

Fig. 14.4 shows the essential geometry of the polar cap. In this model the emitting region is limited by those magnetic field lines that become tangential to the velocity of light cylinder, i.e. those that bound the open lines of the polar cap.

Assuming a simple dipole field as in the aligned rotator, we can find the angular radius of the polar cap by tracing the limiting field lines back to the surface. An individual field line follows polar coordinates r, θ according to:

$$\left(\frac{\sin \theta}{\sin \theta_s}\right)^2 = \frac{r}{a} \tag{14.4}$$

cutting the surface at a, θ_s.

The field line tangential to the velocity of light cylinder, at $r = c/\Omega$, therefore has

$$\sin \theta_s = \left(\frac{\Omega a}{c}\right)^{1/2}. \tag{14.5}$$

At the edge of the polar cap (coordinates θ_0, a), this field line is inclined to the radius vector by $\tan^{-1}(\frac{1}{2}\tan\theta_0)$, so that the field line is at a polar angle δ_s given by

$$\delta_s = \theta_0 + \tan^{-1}(\tfrac{1}{2}\tan \theta_0). \tag{14.6}$$

Since δ_0 and θ_0 are in practice both small, we obtain the two results:

(i) the diameter of the polar cap, in magnetic latitude, for a pulsar with radius a, is

$$2\theta_o = 2\left(\frac{2\pi a}{c}\right)^{1/2} P^{-1/2}; \tag{14.7}$$

(ii) the angular width of a beam determined by the cone of limiting field lines at the surface is

$$2\delta_s = 3\theta_0. \tag{14.8}$$

For an orthogonal rotator, with $a = 15$ km and period P seconds, we find an expected beamwidth for a source at the pulsar surface

$$2\delta_s = 3P^{-1/2} \text{ degrees.}$$

According to this model, the widths for pulsars with periods ranging from 100 milliseconds to 4 seconds should range from 30° to 1.5°. This simple relationship is not obvious in the observed data on pulse widths, but as we

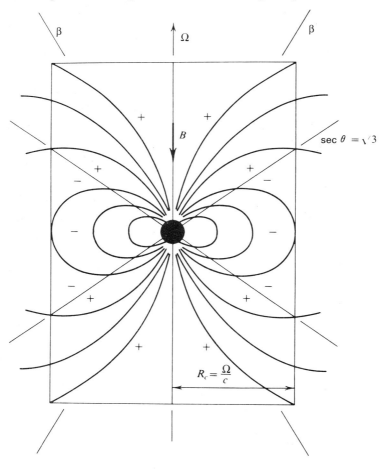

Fig. 14.4. The geometry of the polar cap. In this simple case of an aligned magnetic dipole, the limiting field lines at angle β are those that become tangential to the velocity of light cylinder. The lines at $\sec\theta = \sqrt{3}$ divide the positive and negative domains of the magnetosphere.

14.4 The polar cap and the source of radio emission

have seen in Chapter 12 a more detailed study including the polarisation characteristics gives a better measure of the true beamwidth than does the intensity alone. We then find that the beamwidth at 400 MHz is close to $13\,P^{-1/2}$ degrees; the beamwidth for a one-second period pulsar is 13° rather than 3°, and the variation with P is slightly different.

This wider beam indicates an origin further out in the diverging field lines of the polar cap. Since the angular width of the cap varies as the square root of the radial distance, the beamwidth indicates a distance of about $20a$ for $P = 1$ second, and $7a$ for $P = 0.1$ second. This distance is less certain for the short period pulsars, since for these we only observe the centre of the full radiation profile. For the millisecond pulsars this analysis places the source at a considerable fraction of the radial distance out to the velocity of light cylinder, so that there may be an appreciable effect from relativistic beaming.

Evidence in favour of a location at such a distance above the surface of the pulsar comes from an observation at 327 MHz of differential scintillation between the the outer components of the profile of PSR 1237+25 (Chapter 17). This suggests that the regions of origin are separated by a distance of order 10^8 cm, i.e. about $30\,a$. Radiation from the curved outer field lines appears to come from an annulus or halo around the star itself; the width of this halo in this case agrees with expectation when the emission region is at the height that we have deduced from the angular width of the integrated profile.

For most individual pulsars there is a clear dependence of the angular width on radio frequency v, in which the width varies as $v^{-0.25}$ at frequencies below about 1 GHz. This was first interpreted in terms of diverging field lines by Komesaroff (1970). Lower radio frequencies originate further out; we can therefore map out the variation of frequency along the nearly radial lines of the cone. The observed frequency dependence of profile width shows that the radius to frequency mapping takes the form $v \propto r^{-2}$. The radial distance of the emitting region for a long period pulsar would then be of order

$$r = 20\,a\,(v_{\mathrm{MHz}}/300)^{-1/2} \tag{14.9}$$

The angular width of the corresponding region at the surface would be about 1°, i.e. about 250 metres across. The variations of excitation, which account for the structure of the integrated profiles and the micropulses, would all occur within this region.

15

Radiation processes

The circumstances under which the pulsars radiate their intense beams of radiation are outside the range of previous experience, and we must admit that many aspects are not yet understood. Starting from the observations of beam shapes and polarisation, we can set out consistent geometric schemes that locate the emitting regions within the pulsar magnetosphere; this is the subject of the next chapter. We find that the high-energy emission typical of the Crab and the Vela Pulsars is located far out in the magnetosphere, close to the velocity-of-light cylinder, while the radio emission of most pulsars, including the precursor pulse of the Crab Pulsar, originates nearer the surface. There is also an obvious difference in the radiation processes between the high-energy and the radio regimes, since the latter must be coherent to produce the very high observed brightness temperatures. The high-energy radiation is very broad-band, which is typical of synchrotron and curvature radiation. The radio regime (apart from the special case of the Crab Pulsar) is narrow-band, which is typical of coherent radio mechanisms. In both regimes, however, the geometry of the emitting region is dominated by the magnetic field.

In this chapter, we set out the standard theory of cyclotron, synchrotron and curvature radiation. The geometry of the emitting regions is discussed in Chapter 14 and in Chapter 16 we discuss the physical conditions in the emitting regions.

15.1 Cyclotron radiation

The radiation from a particle with charge e and mass m, moving with non-relativistic velocity on a circular path in a magnetic field B is at the Larmor frequency

$$\nu_L = \frac{eB}{mc}. \tag{15.1}$$

15.1 Cyclotron radiation

For an electron or positron, $v_L = 2.8$ MHz gauss^{-1}.

The rate of loss of energy through radiation is

$$-\frac{dE}{dt} = \frac{8\pi^2}{3} \frac{v_L^2 \beta^2 e^2}{c} \tag{15.2}$$

where the velocity of the charged particle is βc. For cyclotron radiation at the typical radio frequency of 100 MHz, the loss rate is $2.03 \times 10^{-19} \beta^2$ W. The polar diagram of the radiation is shown in Fig. 15.1(a). Here I is the intensity; Q and V are the linearly and circularly polarised components respectively.

When the velocity βc reaches relativistic values, the gyrofrequency v_g is reduced below the value v_L given by equation 15.1 because of the increased mass of the electron; however, the relativistic velocity also leads to the radiation of harmonics of the gyrofrequency. At high relativistic energies this becomes the synchrotron radiation discussed in the next section. But even when most of the radiated power is in the harmonics, the fundamental cyclotron radiation is still emitted with the intensity given by equation 15.2, with the actual gyrofrequency substituted for v_L. Although for an individual electron this radiation may now be insignificant in comparison with the harmonic radiation, the coherent radiation from a bunch of electrons may be concentrated in the fundamental and the lower

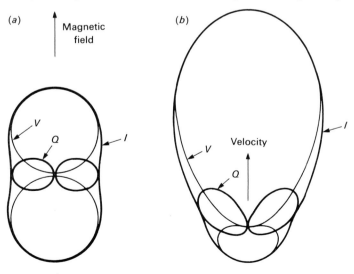

Fig. 15.1. The radiation pattern of cyclotron radiation (a) from an electron in a circular orbit perpendicular to a magnetic field, (b) from an electron streaming along the magnetic field. I, Q, V are Stokes parameters (after Epstein 1973).

harmonics only. We later consider the possibility that the radiation in the radio regime is the coherent cyclotron radiation from bunches of charged particles.

In the pulsar magnetosphere the motion of charged particles must be a combination of gyration about the field lines and streaming along the field lines. If the pitch angle ψ, which is the angle between the velocity vector and the field vector, is small compared with Γ^{-1}, where Γ is the relativistic factor of the particles, i.e.

$$\Gamma = (1-\beta^2)^{-1/2}$$

the radiation can be treated as gyro-radiation within a moving reference frame. The streaming velocity then results in a Doppler shift, and in a beaming effect which concentrates the radiation in the direction of streaming (Fig. 15.1(b)).

The more general case for larger pitch angles can be considered as part of the theory of synchrotron radiation.

15.2 Synchrotron radiation

A charged particle gyrating with high velocity ($\beta \approx 1$) radiates a spectrum of harmonics which extends to frequencies of order $\Gamma^2 \nu_L$, i.e. to Γ^3 times the gyrofrequency ν_g. When Γ is large, this radiation may be regarded as a continuous spectrum. This radiation is synchrotron radiation, also known as magnetobremsstrahlung.

The theory of synchrotron radiation has been presented in detail by Ginzburg and Syrovatskii (1965, 1969). A simple understanding of the essential points may be obtained by considering the electric field radiated by a single electron, gyrating perpendicular to the magnetic field, and observed in the plane of its orbit. This field consists of pulses each occurring as the electron travels towards the observer. The relativistic velocity of the electron concentrates the field in the forward direction, into an angle Γ^{-1}. The observer therefore sees a pulse while the electron traverses an arc Γ^{-1} of its orbit, which occurs in time $\Gamma^{-1}\nu_g^{-1} = \nu_L^{-1}$. Since the electron is travelling towards the observer at this time the duration of the pulse is further compressed by a factor $(1-\beta)$, with the overall result that the pulse width is approximately $\Gamma^{-2}\nu_L^{-1} = \Gamma^{-3}\nu_g$. Harmonics of the gyrofrequency are therefore emitted up to a factor Γ^3, or up to Γ^2 of the Larmor frequency. The full theory defines a 'critical frequency', which is close to this high harmonic frequency.

The radiation is concentrated in the plane of gyration, within an angle Γ^{-1} approximately. (The low harmonics are less concentrated in angle; the nth harmonic is concentrated roughly within an angle $n^{-1/3}$ radian). For

15.2 Synchrotron radiation

particles streaming along a magnetic field line, with pitch angle θ, the radiation is concentrated in a cone with half-angle θ, i.e. the radiation is concentrated in the direction of the electron velocity. The radiation is then determined by the component of the field perpendicular to the electron velocity, so that we may write $B_\perp$ in formulae for critical frequency and intensity.

The polarisation of the radiation may be understood from a description of the electron acceleration as seen by the observer. For the simplest case of the observer in the plane of gyration, with no streaming motion, the polarisation is plane (Fig. 15.2(a)). An observer above or below the plane will see a circular component (Fig. 15.2(b)), giving elliptical radiation, whose ellipticity diminishes to zero at edges of the beam where the polarisation approaches pure circular (Fig. 15.2(c)). A cross-section across the cone of radiation from an electron with a large component of velocity along the field line shows similar polarisation characteristics.

The spectrum of synchrotron radiation, expressed as the total power radiated in all directions, is shown in Fig. 15.3. Below the frequency v_m, where the power is a maximum, the spectrum is a power law, proportional to $v^{1/3}$; above v_m it falls exponentially as $\exp[-(v/v_c)]$.

The essential characteristics of the radiated power are as follows:
Frequency (MHz) in field B (gauss), for particle of energy E

$$\begin{aligned} v_m &= 1.2 B\, \Gamma^2 \\ &= 1.8 \times 10^{12}\, B\, (E_{\text{erg}})^2 \\ &= 4.6 B\, (E_{\text{MeV}})^2. \end{aligned} \qquad (15.3)$$

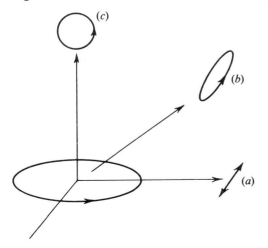

Fig. 15.2. Polarisation of radiation from an electron in a circular orbit.

Total radiated power (spectral density) at ν_m

$$p(\nu_m) = 2.16 \times 10^{-22} B \text{ erg s}^{-1} \text{ Hz}^{-1} \qquad (15.4)$$

Intensity at frequency ν (approximately)

$$p(\nu) \approx 4 \times 10^{-22} B \left(\frac{\nu}{\nu_c}\right)^{1/3} \exp\left(-\frac{\nu}{\nu_c}\right) \text{ erg s}^{-1} \text{ Hz}^{-1} \text{ sr}^{-1} \qquad (15.5)$$

Lifetime of electrons (time to lose half energy by synchrotron radiation)

$$T_s = \frac{5.1 \times 10^8}{\Gamma B^2} \text{ s}. \qquad (15.6)$$

The polarisation at frequencies near ν_m is shown in Fig. 15.4 which represents a cross-section through the fan beam of radiation from a particle moving in an orbit perpendicular to the magnetic field. The polarisation is presented as a linear component $(Q^2+U^2)^{1/2}$ and a circular component V, where I, Q, U, V are the Stokes parameters. The beamwidth is approximately Γ^{-1} at frequency ν_m and above; it increases at lower frequencies where it equals approximately $\Gamma^{-1}(\nu_c/\nu)^{1/3}$.

It is important to note that the circular polarisation changes hand across the beam. Accordingly, if an ensemble of electrons with a range of pitch

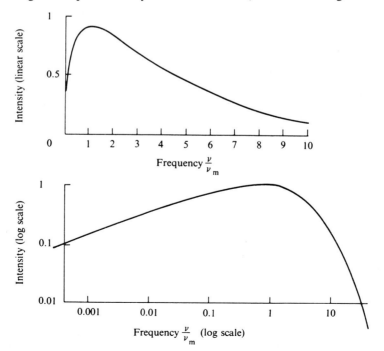

Fig. 15.3. Spectrum of synchrotron radiation, on linear and logarithmic scales.

angles exceeding Γ^{-1} contributes to the radiation, so that the beamwidth of their combined radiation exceeds the width of the beam in Fig. 15.4, the average value of the circular component will be diminished. In practical astrophysical situations, this may remove the circular component entirely, leaving only a linear polarisation amounting to about 70% at most.

15.3 Curvature radiation

In the very high magnetic field of the pulsar magnetosphere, an electron may follow the path of a magnetic field line very closely, with pitch angle nearly zero. The field line will generally be curved, so that the electron will be accelerated transversely and will radiate. The radiation, which is closely related to synchrotron radiation, is called curvature radiation.

An electron with relativistic velocity $v \approx c$, constrained to follow a path with radius of curvature, ϱ, radiates in a similar way to an electron in a circular orbit with gyrofrequency $c/2\pi\varrho$. As in synchrotron radiation, there is a critical frequency given approximately by

$$v_c \approx \frac{c}{2\pi\varrho} \Gamma^3 \qquad (15.7)$$

The maximum intensity from an electron with energy E_e (eV) is at a frequency $v_m \approx 10^{-6} E_e^3 \varrho^{-1}$. The spectrum is of the same form as synchrotron

Fig. 15.4. Angular distribution of synchrotron radiation from a collimated monoenergetic electron beam. ———— total power (I)) —·—·—·— circular component (V); — — — — linear component $(Q^2 + U^2)^{1/2}$; the angular scale is in units of Γ^{-1}.

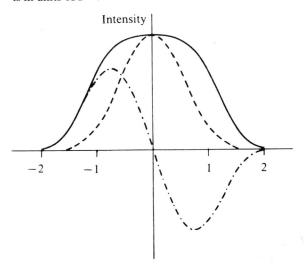

radiation, varying as $v^{1/3}$ at low frequencies and falling exponentially above v_c. The intensity depends on Γ instead of B; at frequencies below v_c it is given by Jackson (1962) as:

$$I(v) \approx \frac{1}{2\pi} \frac{e^2}{c} \frac{c}{\varrho} \Gamma \left(\frac{v}{v_c}\right)^{1/3}. \tag{15.8}$$

The total radiated power is $10^{-31} E_e^4 \varrho^{-2}$ erg s^{-1}.

As a numerical example, we follow the suggestion by Komesaroff (1970) that pulsar radio emission is due to curvature radiation from particles streaming along polar field lines. The ratio ϱ/c is of the order of the pulsar period ($\sim$1 s) and v_c must be at high radio frequencies ($\sim$1 GHz), giving $\gamma \approx 10^3$. For these values of curvature and particle energy, the intensity from a single electron at frequencies near v_c is of order 10^{-27} erg s^{-1} Hz^{-1}. This is very much smaller than the cyclotron or synchrotron radiated intensity; however, the intensity is in practice determined mainly by coherence conditions, and intensity alone does not provide sufficient grounds for a choice of radiation mechanisms. It is, in any case, possible that much higher values of Γ are involved in the curvature radiation.

The polarisation of curvature radiation is linear in the plane of the orbit. On either side the polarisation approaches circular at the edge of the radiated beam. This is exactly the same as for synchrotron radiation.

15.4 The synchrotron spectrum

We have so far considered the synchrotron and curvature radiation from a single monoenergetic charged particle. In practice there will be an ensemble of electrons or positrons with a range of energies. The radiation from each is concentrated around its critical frequency, so that the resultant spectrum depends on the distribution of critical frequency among the ensemble.

If the particle energies are distributed according to a power law with index γ so that

$$N(E) \propto E^{-\gamma} \tag{15.9}$$

then the spectrum also follows a power law:

$$P(v) \propto v^{-\alpha} \tag{15.10}$$

where the spectral index $\alpha = (\gamma-1)/2$. This applies only if the energy power law extends over a sufficient range of energies. If there is a change of exponent γ in the energy spectrum, this will be reflected in a change of exponent in the radiation spectrum, but the change will be smoothed over a range of frequencies. In particular, if there is a cut-off at low energies in the particle spectrum, the radiation spectrum will not cut off abruptly at the corresponding critical frequency, but decline slowly into the $\nu^{1/3}$ power law spectrum of single particles below the critical frequency. A full analysis is given by Ginzburg and Syrovatsky (1969).

In curvature radiation the relation between particle energy spectrum and radiated spectrum is given by

$$\gamma = 3\alpha + 1 \text{ (curvature)} \tag{15.11}$$

instead of

$$\gamma = 2\alpha + 1 \text{ (synchrotron)}. \tag{15.12}$$

15.5 Self absorption

The radiation from an ensemble of energetic charged particles, radiating independently, may in some sense be regarded as thermal radiation. Corresponding to the synchrotron emission process there is, as in any thermal process, an absorption process whose importance grows with an increasing population of radiating particles along the line of sight. The brightness temperature of the radiation at any particular frequency from the total population cannot exceed the temperature corresponding to the energy of those particles that radiate principally near that frequency; when the intensity approaches this level, as the radiation passes along the line of sight, absorption is nearly balancing emission.

Self-absorption is observed in many celestial synchrotron sources at low radio frequencies. It has been proposed that it occurs in the infrared spectrum of the Crab Pulsar; we consider and reject this possibility in Chapter 16.

This discussion of self-absorption does not apply in the radio frequency radiation from any pulsar. Here the observed brightness temperatures are many orders of magnitude higher than any conceivable equivalent particle energies, and the radiation must be coherent.

15.6 Inverse Compton radiation

In the classical Compton effect, high-energy photons collide with low-energy electrons; the photons lose energy to the electrons and emerge with longer wavelength. The inverse effect involves the transfer of energy to radiation from high-energy electrons. It may be regarded as synchrotron radiation from the high-energy electrons as they move in the electromagnetic field of the radiation.

Inverse Compton radiation from a cloud of high-energy electrons therefore increases the total flux of radiation energy and puts the increased energy in shorter wavelengths. It is therefore a radiation mechanism for high-energy particles which does not depend on collisions or on a steady magnetic field.

It has been suggested that the gamma-ray radiation from the Crab and Vela Pulsars is inverse Compton radiation from high energy particles high in the magnetosphere. These particles would be acting on radio or other low frequency radiation arising from nearer the pulsar surface. There are two arguments against this hypothesis. First, it is difficult to find a geometric model that will fit the observed pulse shapes, while a synchrotron or curvature model depending solely on a steady magnetic field achieves this very simply (Chapter 14). Second, the efficiency of the process is smaller than that for synchrotron or curvature radiation, provided that the steady magnetic field in the radiating region is as large as is currently accepted.

15.7 Maser amplification

The very high brightness temperatures observed in radio emission from pulsars prompted suggestions, notably from Ginzburg and Zheleznyakov (1970), that some form of maser amplification might occur outside the actual source of the radiation. The process is well known for radio-spectral line radiation from water vapour and the hydroxyl radical in molecular clouds. It depends upon a non-thermal energy spectrum in the amplifying region; for line radiation this is an inversion of populations between energy levels, while for continuum radiation it must consist of a division between a very high energy component and a low-energy component.

The maser process is a coherent amplification, which occurs in a narrow band of wavelengths, and it is hard to see how this can be applied over a very wide bandwidth. It is also very difficult for a maser amplification theory to explain the detailed characteristics of polarisation in the radio pulses.

15.8 Coherence in the radio emission

Since no method of amplification outside the radio source can account for the observed very high brightness temperatures, we are forced to conclude that the radiation is coherent. Any radiation mechanism is made more efficient if electrons are constrained to move in synchronism, so that a group of N electrons moves like a single particle with charge Ne. The radiated power is then proportional to N^2 rather than N. Furthermore, the electrons are obviously not in random thermal motion, and the brightness of the source is not limited by self-absorption.

Coherent motion is only effective if the radiation from all electrons in the bunch is in phase. For isotropic, or nearly isotropic, radiation the requirement is that the bunches of particles have a scale of less than half a wavelength, λ. If the radiation is beamed, then a flat disc-like bunch is an effective coherent radiator, provided that the axis of the disc is directed parallel to the radiation. Along the axis of the disc the thickness must be less than half a wavelength, as before; but the width may be very much greater. For the extreme case of a smooth but slightly curved disc with radius of curvature equal to the pulsar radius R, as might be found over a polar cap, the radius of the coherent zone could be as large as $(R\lambda)^{1/2}$.

The formation of such bunches depends on instabilities in the electron flow. An example is given by Goldreich and Keeley (1971), who investigated the stability of mono-energetic particles constrained to move in a ring. By evaluating the interactions between particles they showed that a perturbation in density may tend to grow. Synchrotron radiation from the bunch will then be enhanced at the longer wavelengths. If the scale of the perturbation, expressed as an angular width round the ring, is small compared with γ^{-3}, then the radiation from the bunch of particles will be in phase over the whole of the synchrotron spectrum, giving enhanced radiation. Larger angular scales of bunching would provide coherence only for the lower harmonics of the synchrotron radiation; in the extreme a perturbation consisting of a single sinusoidal distribution round the ring would only enhance the fundamental gyrofrequency radiation.

As we have seen in Section 15.3, curvature radiation is very similar to synchrotron radiation, so that coherence in curvature radiation has a similar effect. If the radio emission is the curvature radiation from electrons constrained to follow field lines from the polar cap, which have a large radius of curvature, the curvature spectrum itself has little significance, and the radiated spectrum must depend almost entirely on the structure of the coherent bunches of electrons.

15.9 Relativistic beaming

We argue in the following chapter that the high-energy radiation from the Crab and Vela Pulsars originates close to the velocity-of-light cylinder, so that the source is moving with an appreciable fraction of the velocity of light. Whatever the detailed radiation mechanism, this high velocity v has a very important effect on the radiation, concentrating it in a beam along the instantaneous direction of motion. If the source radiated isotropically, the observer would see the radiation concentrated in a beam with width approximately equal to Γ^{-1}, where the relativistic factor Γ

is related to the velocity of the source rather than to the energy of the radiating particles, i.e.

$$\Gamma = \left(1 - \frac{v^2}{c^2}\right)^{-1/2}. \tag{15.13}$$

For the rotating pulsar, there is a further effect due to time compression by a factor Γ^2 when the source is moving towards the observer, so that the beam sweeps across the observer in time τ given approximately by

$$2\pi \frac{\tau}{P} \approx \frac{1}{2\Gamma^3}. \tag{15.14}$$

There is also a Doppler shift of frequency. When the source is moving directly towards the observer the frequency is increased by a factor

$$v'/v = \left(\frac{1 + \frac{v}{c}}{1 - \frac{v}{c}}\right)^{1/2}. \tag{15.15}$$

This factor changes within the pulse duration, so that any narrow spectral feature (such as cyclotron resonance) would be effectively smeared out.

During the pulse, when the source is moving towards the observer, the intensity is increased: outside the pulse, the intensity is decreased. The factors may be large; they depend on the spectrum of the emitted radiation, because of the Doppler shift. Calculations may be found in Smith (1970, 1971) and Zheleznyakov (1971).

16

The pulsar emission mechanisms

As we have seen in Chapter 14, there are two emitting regions to consider: the radio emitter in the polar cap region, and the high-energy emitter (optical, X-ray, gamma-ray) in the outer magnetosphere gap. These are located at or close to the two possible vacuum gaps in which charged particles are accelerated to high energy. In both regions the radiation is directed along the magnetic field lines, i.e. along the direction of particle flow. In both, the *angular* spread of the radiation, which is seen as an integrated pulse shape, is primarily attributed to a *spatial* distribution of radiating sources.

The high-energy radiation from the young pulsars has a smooth and very broad spectrum, and no short-term fluctuations in intensity. It therefore has the characteristics of classical synchrotron or curvature radiation. The typical radio emission from a polar cap also extends over a wide frequency range, but in contrast to the high-energy radiation most characteristics of pulse shape and intensity in the radio regime are relatively narrow-band. Further, the intensity is very much higher than is allowable in incoherent radiation processes, and it is also very variable. The coherent mechanism, which is evidently inherent in the radio emission, is less well understood than the process of the high-energy emission in the outer gaps, and we therefore start with a simple analysis of the latter.

16.1 The outer magnetosphere gaps

The outer gaps are populated with streams of charged particles and radiation, both well collimated along the direction of the magnetic field lines. Fig. 16.1 shows a section of an outer gap.

The creation of an electron-positron pair (at point 2 in the diagram) results from the interaction of a gamma-ray with either the magnetic field or, more probably, a lower energy photon. The pair are separately accelerated along the field lines, reaching energies comparable with the maximum available potential, i.e. about 10^{14} volts for the Crab and Vela pulsars. They radiate gamma-rays, either by curvature radiation or by inverse Compton collisions with low energy photons. These gamma-rays then can create further electron–positron pairs. The effect of the cascade process is to create streams of high-energy particles, i.e. positrons and electrons, one streaming inward and the other outward along the magnetic field lines. Successive stages of the cascade drift across the gap boundary towards the poles, allowing replenishment of the charged particles in the plasma over the magnetic poles. The importance of this cascade process was first pointed out by Sturrock in 1971.

Many different interactions between particles and photons are possible in this cascade process. Radiation from the oppositely moving electrons and positrons forms crossing beams of photons and particles. Low-energy particles beyond the gap boundary can radiate infrared photons back into the gap. The high-energy particles in the gap may radiate through interaction with the magnetic field (synchrotron or curvature radiation) or by inverse Compton collisions with low-energy photons. Some idea of the balance between these processes can be obtained from a paper by Cheng, Ho and Ruderman (1981). In this chapter we prefer to deduce as much as

Fig. 16.1. The cascade process in the outer magnetospheric gap. Electrons (−) and positrons (+) accelerated in the gap emit gamma-rays (∿∿), which, in turn, create pairs of electrons and positrons, the process moving progressively into the polar cap region.

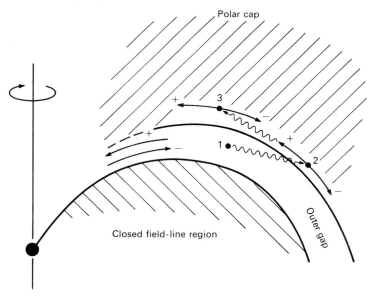

possible directly from the observed characteristics of the radiation from the Crab and Vela Pulsars. Fortunately, theory and observation now agree well for the high energy emission from the outer gap region.

16.3 The Crab Pulsar: high energy emission

The outer magnetosphere gap may extend over a large radial distance and it may have a thickness of several kilometres. We may therefore expect the energy spectrum of the radiating particles to be different in different parts of the gap; the effect would be observable as differences in spectrum in different parts of the observed pulse profile. The only differences observed are:
  (i) in the X-ray region there is excess radiation in the bridge between the two pulses and in the second pulse (Section 7.1.2);
  (ii) at the very highest energy gamma-rays (around 10^{14} eV) the pulses are narrow (resembling the narrow radio pulses).

Apart from these, there is a close similarity in pulse shape over more than 10 decades of photon energy; it is inescapable that over the whole of this range the source is the same and that it contains the same energy spectrum of particles, radiating by the same mechanism. We deduce that the cascade of processes in the gap produces a continuous energy spectrum of electrons and positrons within the whole of the radiating region. The two spectral features referred to above suggest that:
  (i) excess X-radiation originates in identifiable regions, possibly at parts of the gap nearer to the star;
  (ii) the highest energy particles, radiating the highest energy gamma rays, are generated only near the inner edge of the gap, where the charge depletion, and consequently the electric field, are maximum.

The observed radiation comes only from the outward flow of charged particles. Gamma-rays from the inward flow cannot traverse the magnetosphere without being absorbed through pair-creation; since the pulse profile is the same over the whole spectrum, it follows that there is no observable radiation from the inward particle flow at lower photon energies.

The high degree of linear polarisation, observed in the visible spectrum but presumably applying over the whole high-energy spectrum, is indicative of curvature radiation. Some previous analyses, starting with Shklovsky (1970) and including more recently Zheleznyakov and Shaposhnikov (1972), Davila *et al.* (1980) and Knight (1982), are based on synchrotron radiation; as we note later, the required physical conditions such as particle energy and density are not very different at the highest energies but important differences exist at lower energies.

The radius of curvature R of the magnetic field lines in the outer gap of the Crab Pulsar is about 10^8 cm. Using the theory of curvature radiation, we now find the relativistic factor Γ of electrons or positrons which radiate the highest energy gamma rays, observed at $\nu = 10^{27}$ Hz. The relativistic Doppler shift due to the relativistic motion of the source for $v/c = 0.9$, as in Section 14.9, is a factor of 4.4. The nominal upper limit to the spectrum radiated by a single electron is, say, five times the critical frequency ν_c, where

$$\nu_c = 7.2 \times 10^9 \, R^{-1} \, \Gamma^3. \tag{16.1}$$

Reducing the observed upper limit by the Doppler factor, we deduce that the relativistic factor for the maximum energy electrons is $\Gamma \approx 5 \times 10^7$, equivalent to 2.5×10^{13} eV.

The lowest frequency in the continuous observed spectrum from the Crab Pulsar is $\nu \approx 10^{14}$ Hz. The spectrum falls very rapidly below this frequency (a point to which we return later). Allowing for a Doppler shift, the critical frequency for the particles responsible for this radiation must be $\nu_c \approx 10^{14-15}$ Hz corresponding to an energy of 10^9 eV. If the particle energy spectrum is a power law with exponent γ, then it is related to the radiation spectral index α by

$$\gamma = 3\alpha + 1 \tag{16.2}$$

The exponent γ therefore varies from 4 at the highest energies to 1 in the optical range. The integrated particle energy is concentrated in the range $\Gamma = 10^4$ to 10^5.

The total power radiated by a single electron or positron is related to ν_c and R by

$$-\frac{dE}{dt} = 3.6 \times 10^{-22} \, \nu_c^{4/3} \, R^{-2/3} \, \text{erg s}^{-1} \, . \tag{16.3}$$

The lifetime of the highest energy particles is therefore about 1 microsecond, during which time they would travel 300 metres. They therefore reach an equilibrium energy of order 5×10^{13} eV, gaining energy from the electric field and losing energy via curvature radiation. The lowest energy particles, with $\Gamma \approx 5 \times 10^4$ and below, radiate very little of their energy in the magnetosphere; the outward stream of these particles provides the energy supply for the Crab Nebula.

The total power radiated by the pulsar in the optical and ultraviolet range may be used to estimate the flux of these lower energy particles,

16.3 The infrared spectrum

which carry most of the energy out to the Nebula. The radiated power is of order 10^{31} erg s^{-1} (Chapter 7). The total power radiated by a single electron is $10^{-8.3} \Gamma^4 R^{-2}$ erg s^{-1}, so that we require 10^{37} electrons with $\Gamma \approx 3 \times 10^4$ to be radiating at any time. These radiate for about 1 millisecond as they pass through the radiating zone; the outward energy flow carried by the particles is therefore of order 10^{38} erg s^{-1}. This may be compared with the total rate of loss of rotational energy ($\sim 10^{40}$ erg s^{-1}) and with the minimum total energy ($\sim 10^{38}$ erg s^{-1}) required to fuel the radiation from the Nebula.

16.3 The infrared spectrum

The spectrum of the pulsed radiation from the Crab Pulsar shows a peak in the optical region, and a steep fall in the infrared. Curvature or synchrotron radiation cannot alone give such a steep fall (Fig. 16.2); the low-frequency spectral index cannot exceed 0.7 even if the energy spectrum of the radiating particles has a sharp cutoff at a limiting low energy. It has therefore been proposed that this steep fall is due to absorption.

Observations of the pulse shape at infrared wavelengths show no differ-

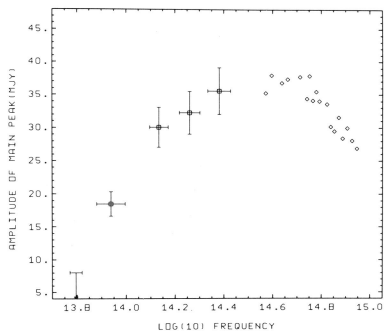

Fig. 16.2. The infrared spectrum of the Crab Pulsar (Middleditch *et al.* 1983). The optical points (plotted as diamonds) are from Oke (1969).

ence from the high-energy regime. Self-absorption within the source should affect the peak intensity first, and a fall in mean intensity by a factor greater than five should produce a very large effect. Absorption outside the pulsar would not affect the pulse shape, but the physical nature of such an absorber has not yet been understood. The infrared spectrum still awaits an explanation.

16.4 The Crab Pulsar double radio pulse

The narrow radio pulses, referred to as the 'main' and 'secondary' or 'interpulse', occur precisely aligned in time with the high-energy pulses. Their widths are similar to those of the highest-energy gamma pulses, and it is a reasonable assumption that they originate in the same high energy charged particles. They are linearly polarised, with the same almost constant position angle in both components. The radio pulses cannot, however, simply be the low frequency extreme of broad-band curvature radiation, since their very high brightness temperature and the very variable pulse intensity indicate that the radiation process is coherent.

The cascade process in the outer magnetosphere gap may lead naturally to bunching of secondary electrons or positrons, through a succession of pair-creation and inverse Compton collisions with photons. The products will all be travelling together at nearly the velocity of light. A bunch of charge will radiate in the forward direction coherently at wavelengths larger than its longitudinal spread; the received radio wavelength will be much reduced due to a very large Doppler shift. The enhanced brightness will, of course, only occur at radio wavelengths, since the coherence is least at shorter wavelengths.

If the radio emission is indeed coherent curvature radiation from particles with energy $\sim 5 \times 10^{13}$ eV, the coherence must increase the brightness by a factor of at least 10^{10}, implying that at least 10^5 electrons are bunched together. The variability in pulse intensity would then represent the variations in the populations of the bunches.

16.5 The Vela Pulsar high-energy pulses

The high-energy pulses from the Vela Pulsar are similar to those from the Crab Pulsar, except that the observable spectrum extends only down to 50 MeV. Below this photon energy, the spectrum falls below the extrapolation with spectral index $=-1$ (Fig. 7.5 Chapter 7). (The double pulse in the optical range has a different shape and spacing, and is not a continuation of the high-energy spectrum.)

There is no information on the polarisation of the high-energy pulses, but it is reasonable to assume that the emission mechanism is the same as for the Crab, i.e. curvature radiation. There must therefore be a reduction in the lower energy end of the particle energy distribution within the gap, so that only particles with $\Gamma \geqslant 10^7$ are contributing to the observable beam.

The optical pulses originate at smaller radial distances; they may represent synchrotron radiation from lower energy particles, but no precise analysis has yet been possible.

16.6 The radio pulses

The geometrical interpretation of the integrated pulse profiles, including their dependence on radio frequency and their polarisation is unequivocal: the radio beam of all pulsars, including the 'pre-cursor' beam of the Crab Pulsar, originates in the magnetosphere directly over a magnetic pole. The radiation is coherent, and each location radiates a definite radio frequency. The radiation is highly polarised, and the polarisation position angle is related to the magnetic field direction at the emitter. Circular polarisation is also observed, often as the major polarised component; it is most likely to be seen in radiation from near the centre of the emitting region.

These observations indicate that the source is resonant, and that the magnetic field is a main determinant of the radio frequency. The simplest hypothesis is that the resonance is a pure gyroresonance at angular frequency $\omega = eH/\Gamma mc$. For electrons in a magnetic field of 10^{12} gauss, as expected near the poles, this requires an impossibly high energy, with $\Gamma > 10^8$ to reduce the cyclotron frequency to below 1 GHz. Particles with such high energy, if they could be generated, would lose energy very rapidly in high energy photons. The observed angular widths of the pulse profiles indicate, however, that the radio emission from long-period pulsars originates at a height of order 40 stellar radii. Here the field is reduced to the order of 10^7 gauss, and electrons with $\Gamma \approx 10^2$ or 10^3 would have a gyro-frequency in the observed range.

The resonance may therefore be determined primarily by the magnetic field; this would presumably account for the observation of circular polarisation, though not obviously for a reversal of hand during the integrated pulse profile. We note also that the radius-to-frequency mapping (Chapter 14) indicates a resonant frequency varying as r^{-2}; the radial decrease of field strength would alone indicate a fall as r^{-3}, but a decrease of particle energy as r^{-1} could easily account for this discrepancy.

We conclude that the radio emission is coherent, from bunches of electrons streaming along the field lines of the polar cap, and that it occurs at or near a resonant location determined by the magnetic field and the plasma density. The linear polarisation is in accordance with curvature radiation, but a full analysis must now take account of a complex and anisotropic refractive index in the region of the emission.

The creation of the bunches themselves is also a matter of speculation. The existence of micropulses suggests that bunches may have a very limited extent. Filippenko and Radhakrishnan (1982) suggest that the acceleration of particles in the polar gap is intermittent, consisting of a rapid sequence of sparks. The velocities of charged particles leaving the gap would vary through the development of each spark. As in a klystron valve, this velocity modulation is transformed into density bunching as the electrons stream along the field lines. The classical two-stream instability might also develop, due either to velocity modulation or inherently in the high-density relativistic particle stream (Beskin *et al.* 1988).

The main pulses of the Crab Pulsar, as contrasted with the precursor, are believed to originate in the outer magnetosphere gap (Chapter 14). It is remarkable that the intensities of two very different sources, located in the polar and the outer gaps, are comparable. It should be noted, however, that their spectra are different, and that the pulse height distributions are different: the 'giant pulses' are observed only within the main pulses.

17

Interstellar scintillation and scattering

Optical scintillation is familiar as the twinkling of stars, and as the shimmer of distant objects seen through a heat haze. At radio wavelengths scintillation is encountered in many different circumstances, because there are many kinds of radio transmission paths which contain the necessary phase irregularities. The solar corona, for example, contains an irregular outflowing gas, which disturbs radio waves passing through it from distant objects to the Earth. The effects of this may be thought of either as refraction or as diffraction; in more general terms the waves are scattered, giving rise to an angular spread of waves and to fluctuations in wave amplitude. Similar effects are observed in the Earth's ionosphere.

At the time of discovery of the pulsars, the known examples of radio scintillation gave a rapid fading pattern, not very different from the visible twinkling stars. The comparatively slow, deep fading of radio signals from the pulsars was an entirely new phenomenon, which was first recognised as a form of scintillation by Lyne and Rickett (1968). The basic analysis of scintillation in terms of random refraction in the interstellar medium was presented by Scheuer (1968). He showed that the fluctuations should have a fairly narrow frequency structure, whose width should depend on the distance of the pulsar. Rickett (1969) obtained experimental proof of this, and showed that fluctuations due to scintillation could be clearly distinguished from the various kinds of fluctuation which are intrinsic to the pulsars. Theory and observation are now both well developed, showing such good agreement that scintillation is now used to investigate the interstellar medium and the velocities of pulsars. Interstellar scattering is also observed as a spread in travel time for pulses, seen as a broadening of sharp features; scattering can also cause scintillation on a long time scale of days or months, due to refraction in large-scale components of ionisation in the interstellar medium.

17.1 A thin-screen model

The simplest model is in practice useful for most of the analysis of pulsar scintillation. In Fig. 17.1 random irregularities of refractive index are found between the source and the observer, but effectively concentrated into a thin screen roughly midway along the propagation path. The irregularities have a typical dimension a, and the screen thickness is D. The average refractive index μ is close to unity, so that a variation $\delta\mu$ extending over a length a changes the phase of a wave by an amount $\delta\phi \approx (2\pi/\lambda)a\,\delta\mu$. Since for μ close to unity

$$\mu - 1 \approx \frac{n_e e^2}{2\pi m v^2} \tag{17.1}$$

where n_e is electron density, the phase change for a wave traversing a single irregularity is

$$\delta\phi = r_e \Delta n_e a \lambda \tag{17.2}$$

where r_e is the classical radius of the electrons (2.82×10^{-13} cm) and Δn_e is the fluctuation in electron density corresponding to $\delta\mu$.

A ray passing through the whole screen encounters D/a irregularities, randomly distributed, so that the difference between the phase perturbations of rays separated laterally by more than a becomes

$$\Delta\phi \approx \left(\frac{D}{a}\right)^{1/2} \delta\phi = D^{1/2} a^{1/2} r_e \Delta n_e \lambda . \tag{17.3}$$

We may now employ either geometric optics to describe the angular spread of rays leaving the screen, or diffraction theory to describe the behaviour of a wave front with phase irregularities $\Delta\phi$ impressed upon it, varying with a lateral scale a. Geometrically, the rays are scattered through an angle

Fig. 17.1. Thin screen model of scintillation.

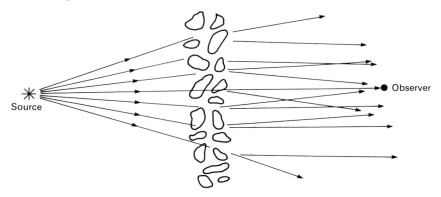

$$\theta_{\text{scat}} \approx \frac{\Delta\phi}{2\pi} \cdot \frac{\lambda}{a} = \frac{1}{2}\left(\frac{D}{a}\right)^{1/2} r_e \Delta n_e \lambda^2 \qquad (17.4)$$

and this, under conditions to be discussed later, is the apparent angular size of the source seen through the screen. Immediately beyond the thin screen there are no amplitude variations across the wavefront. At increasing distance from the screen the amplitude variations build up through interference between rays from various parts of the corrugated wavefront. Separate rays reaching the observer from different parts of the screen will interfere if they have different values of $\Delta\phi$, i.e. if they come from points at least a distance a apart. Hence if the distance from source to observer is L, and

$$L\theta_{\text{scat}} > a \qquad (17.5)$$

a randomly variable amplitude will be observed as the source, or the screen, moves across the line of sight. The ray paths will differ by $\frac{1}{2}\theta_{\text{scat}}^2 L$, which may be a difference of many wavelengths. Interference between rays will therefore differ at different wavelengths; the amplitude will therefore vary over a wavelength difference $\Delta\lambda$ where

$$\frac{\Delta\lambda}{\lambda} \approx \frac{2\lambda}{\theta_{\text{scat}}^2 L}. \qquad (17.6)$$

In practice, the screen is extended along the line of sight so that $L \approx D$. The wavelength range $\Delta\lambda$ is more conveniently expressed as a frequency difference B_s, which for the simple model is given by

$$B_s \approx \frac{8\pi^2 ac}{D^2 (\Delta n_e)^2 r_e^2 \lambda^4}. \qquad (17.7)$$

These simple formulae contain the essential features of the scintillation phenomenon as observed. For example, the apparent angular size of pulsars can be measured by interferometry at long wavelengths, when it is confirmed that $\theta_{\text{scat}} \propto \lambda^2$, while Rickett (1969) found $B_s \propto \lambda^{-4}$ approximately, and decreasing with increasing dispersion measure, as expected if $B_s \propto D^{-2}(\Delta n_e)^{-2}$ and Δn_e was connected with the ionised gas along the line of sight.

17.2 Diffraction theory of scintillation

The wave front leaving the screen of Fig. 17.1 may be treated by diffraction theory. A wave front with randomly distributed irregularities of phase $\Delta\phi$ may be constructed from a plane wave with constant phase, to which is added a range of other plane waves with a distribution of wave normals, forming an angular spectrum of plane waves (Fig. 17.2). These are the scattered waves, covering an angle of $(2\pi/\lambda)a\Delta\phi$. The amplitude of the waves increases with $\Delta\phi$ until $\Delta\phi$ becomes large compared with 1 radian. The interference of these waves at the observation point is then responsible for the scintillation.

This diffraction analysis is particularly valuable in scintillation theory applying to screens which are extended along the line of sight (Section 17.3). Three simple results emerge, however, which apply equally well to the thin screen model:

(i) If $\Delta\phi \ll 1$, which is known as weak scattering, there is a large unscattered plane wave. A point source therefore appears to the observer as a point surrounded by a weak scattered halo. As $\Delta\phi$ increases, the point becomes weaker and the halo becomes stronger.

(ii) The lateral scale of the amplitude irregularities in the wave front is the same as the scale a of the phase irregularities provided that $\Delta\phi \ll 1$. When $\Delta\phi \gg 1$, i.e. for strong scattering, the scale is reduced and the typical scale S of the pattern becomes

$$S = \frac{a}{\Delta\phi} \approx \frac{\lambda}{2\pi\theta_{\text{scat}}}. \tag{17.8}$$

This scale may be observable directly by making simultaneous observations at widely spaced sites, or indirectly by observing the time scale τ of fluctuations at a single site and assuming that the scintillation pattern moves past the observer with a definite velocity V so that $S = V\tau$.

(iii) The amplitude of the wave at the observer is found from the addition of the scattered and unscattered waves with random phases. For

Fig. 17.2. A wavefront W, scattered at a diffracting screen, emerges as a range of wavefronts W_s with a distribution of wave normals.

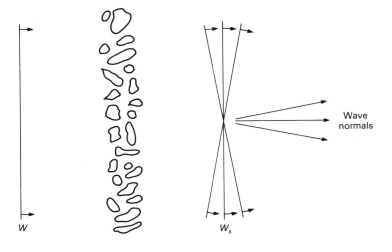

strong scattering there is no unscattered wave and the amplitude distribution approximates to a Rayleigh distribution. The addition of an unscattered wave changes this to a Rice distribution. (A Rayleigh distribution of amplitude corresponds to an exponential distribution of intensity. The Rice distribution is discussed in connection with interplanetary scintillation by Cohen *et al.* (1967).)

17.3 Thick (extended) scattering screen

The thin screen analysis which has been used so far does not depend critically on the location of the screen provided that it is not very close either to the source or the observer. It is a reasonable deduction that a thick screen, occupying most of the space between the source and the observer, will behave similarly to a thin screen. The phase and amplitude modulations will, however, build up progressively, and in the early part of the path there will be an unscattered component which decays exponentially with distance z along the path, so that its relative intensity can be expressed as $\exp(-\beta z)$. Here β is the coefficient of total scattering, given by Uscinski (1968) as:

$$\beta = \pi^{1/2} r_e^2 (\Delta n_e)^2 \, a \, \lambda^2. \tag{17.9}$$

This refers to irregularities concentrated at scale a, which may be defined as the radial distance at which the autocorrelation function of Δn_e falls to $1/e$. Weak scattering corresponds to $\beta z \ll 1$, and strong scattering to $\beta z \gg 1$. Equation 17.9 must in practice be modified to take account of the actual distribution in size of the irregularities, as in the next section. Following the simple case, the angular spectrum of plane waves emerging from a path z has a half-width θ_s (to amplitude $1/e$) given by

$$\theta_s = \frac{\lambda}{\pi a} \text{ for } \beta z \ll 1 \tag{17.10}$$

$$\theta_s = \frac{\lambda}{\pi a} (\beta z)^{1/2} \text{ for } \beta z \gg 1. \tag{17.11}$$

The full analysis for a Kolmogorov distribution of irregularities (see Section 17.5 below) is given by Lee (1976), and a more general review is given by Rickett (1977).

17.4 The Fresnel distance

Close to the thin screen shown in Fig. 17.1, there will be only variations of phase across the wavefront, since amplitude variations occur only beyond a distance where rays from different parts of the screen can cross. The transition to full scintillation is like the classical transition from

Fresnel to Fraunhofer diffraction, and the distance at which it occurs is referred to as the Fresnel distance z_0. It depends only on the scale size a and the wavelength λ, as

$$z_o = \frac{\pi a^2}{2\lambda}. \tag{17.12}$$

Since the Fresnel distance depends on a, it may happen that for a given path only the small scale irregularities will give rise to amplitude scintillation, while the larger scale irregularities are responsible only for refractive effects which we consider in Section 17.13.

17.5 Distribution in size of the irregularities

Although the origin and the development of the fluctuations in electron density are not understood, it is reasonable to expect that they behave like other forms of turbulence, with energy fed into large scale irregularities which degrade into smaller eddies. In such cases, provided that energy is fed in only at a large scale and is lost only at a small scale, the spectrum of irregularities is expected to follow a Kolmogorov power law. If we now define the spectrum $P(q)$ in terms of wave number q, defined by

$$P(q) = \int d^3r \langle n_e(x)n_e(x+r)\rangle \exp(iqr), \tag{17.13}$$

such a law has the form

$$P(q) = C_N^2 \, q^{-\alpha} \tag{17.14}$$

where C_N^2 is a measure of the mean square fluctuation. The power index $\alpha = 11/3$ for Kolmogorov turbulence; we will see that this corresponds well with observation over a wide range of scale sizes.

Scintillation results from a limited range of scale sizes $a = 2\pi q^{-1}$. Fig. 17.3 shows the angular scattering due to different scales of irregularity. Below and left of line A the total phase deviation is less than a radian, and only weak scintillation can occur. Above and to the right of B, interference between the rays does not develop within the propagation path, as it is less than the Fresnel distance. The two limits depend on the wavelength of observation.

The nearly horizontal line in this figure represents a Kolmogorov spectrum. The total angular scattering is found by integration over the relevant range of scale sizes. For longer radio wavelengths the range increases, while at a sufficiently short wavelength the range contracts to the point A. Provided that $\alpha < 4$, the most important contributions to the integral come from the smaller scale sizes. As we will see later, the scale sizes larger than a_2 cause an observable refraction effect, which is responsible for intensity variations on a much longer time scale.

17.6 Dynamic spectra

Fig. 17.4 shows the scintillation of a pulsar observed over a range of radio frequencies. At any one frequency a random fading pattern is seen with a time scale τ, while at any one time the scintillation extends over a typical bandwidth B. Both τ and B may be obtained from such a dynamic spectrum by an autocorrelation analysis. Typical forms of the autocorrelation in time and frequency are shown in Fig. 17.5. The variations in time are closely Gaussian in form, but the variations in frequency depend on the distribution in size of the irregularities, as indicated in the previous section; the form of the autocorrelation function has been used by Wolsczan (1982) and by Roberts and Ables (1982) to confirm the Kolmogorov spectrum.

The time scale τ is usually measured as the time lag where the autocorrelation falls to e^{-1}, while B is usually measured to a frequency separation where the autocorrelation falls to ½.

17.7 Frequency drifting

The dynamic spectrum often shows organised structure, in which the features drift regularly across the observed frequency band, as in Fig. 17.6. This is the effect of refraction in large-scale irregularities. Refraction

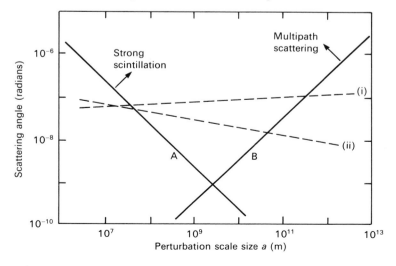

Fig. 17.3. The range of scale size responsible for scintillation as a function of wavelength. To the left of the line A (where $\varphi = \pi$) the total phase deviation is less than one radian; to the right of the line B (where $z\theta = a$) the propagation path is less than the Fresnel distance. The dashed line (i) represents an interstellar medium with power low index $\alpha > 4$, and the line (ii) represents $\alpha < 4$. In this example the source is placed at distance $z = 0.1$ kpc.

218 *Interstellar scintillation and scattering*

moves the whole scintillation pattern laterally; the refraction is greater at lower frequencies. As the pattern moves past the observer, this dispersive effect is seen as a frequency drift of scintillation features.

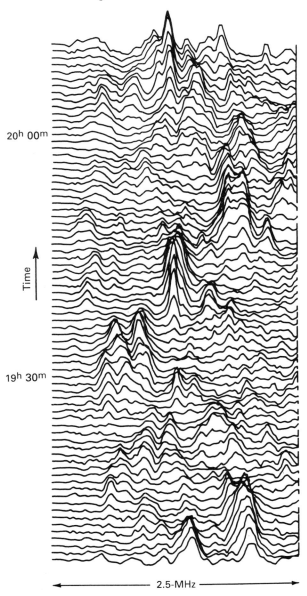

Fig. 17.4. Scintillation of PSR 0329+54 observed simultaneously over a range of radio frequencies.

17.7 Frequency drifting

Fig. 17.5. Autocorrelation of scintillation in time and in frequency: the two-dimensional autocorrelation function of the data in Fig. 17.4. Sections along the time and frequency axes are shown.

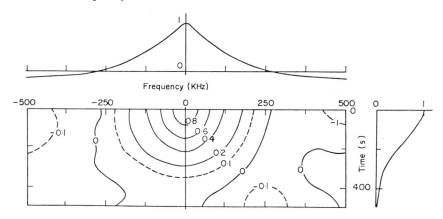

Fig. 17.6. Frequency drifting in the scintillation spectrum of PSR 0823+26 (after Cordes *et al.* 1985).

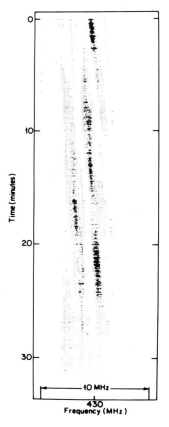

The rate of frequency drifting $d\nu/dt$ is expected to be related to the velocity V of the pattern, the angular spread of rays θ_r and the distance z of the pulsar by:

$$\frac{d\nu}{dt} = \frac{\nu V}{\theta_r z} \sec \psi . \tag{17.15}$$

The term $\sec \psi$ takes account of the angle ψ between the velocity V and the direction of dispersion. This relation has been shown to hold for over 30 pulsars, and it can be used to find θ_r. Since this angle refers to refraction in large scale irregularities (to the right of the line marked $\theta z = a$ in Fig. 17.3), measurements of drift rate allow the spectrum of the irregularities to be extended over several orders of magnitude.

A comparison between the scattering angle θ_s, obtained from the bandwidth and timescale of normal scintillation, and the refraction angle θ_r, shows that the Kolmogorov spectrum of irregularities applies over a range of sizes from 10^9 m to 10^{12} m (Smith and Wright 1985).

Drifting patterns in dynamic spectra often contain organised structures on smaller scales of time and bandwidth. These resemble interference fringes, and they are evidently due to the interference between a small number of predominant rays from the refractive regime. The phenomenon is very variable, but when it is present it indicates the existence of large scale irregularities which may be more intense than those expected from the Kolmogorov law (Hewish 1980).

A further test of the spectrum is available through a detailed study of the variation with observing wavelength of the bandwidth and time scale of scintillation. As the wavelength changes, so does the range of scale sizes responsible for scintillation, and the simple laws of equations 17.4 and 17.7 are modified. Lee and Jokipii (1976) contrasted a model in which sizes are concentrated round a single value, being spread only by a Gaussian distribution, and the model power law distribution with index α. They found that instead of $\theta_s \propto \nu^{-2}$, as in equation 17.4, the power law gave $\theta_s \propto \nu^{-\alpha/(\alpha-2)}$. Hence, for $\alpha = 11/3$ the following differences in frequency dependence may be observed:

		Gaussian	Kolmogorov
Scattering angle	θ_s	-2	-2.2
Bandwidth	B_s	$+4$	$+4.4$
Time scale	τ	$+1$	$+1.2$

17.8 The velocity of the scintillation pattern

The fluctuations of intensity due to scintillation may represent either a random turbulence in the scattering medium, which changes the configuration of the scintillation pattern over a plane containing the observer, or a relative velocity within the system of the observer, medium, and source, so that the pattern drifts past the observer. (The two possibilities are familiar in the patterns of sunlight refracted on to the bottom of a swimming pool, when the waves on the surface are both travelling and changing shape.) Most of pulsar scintillation is found to be due to pattern movement rather than pattern instability, so that the rate of the intensity changes is related to the pattern scale and the drift velocity. For example, a typical drift velocity might be 100 km s^{-1}, with a fading time scale of 10 min; the scale of an unchanging pattern would then be about 60 000 km, larger than the diameter of the Earth.

A drifting and unchanging scintillation pattern was observed by Galt and Lyne (1972) for PSR 0329+54. Observations were made at 408 MHz simultaneously at Jodrell Bank and at Penticton, Canada, 6833 km apart. The orientation of this baseline changes as the Earth rotates; the pulsar is circumpolar so that a complete rotation of the baseline occurred during 24 hours of observation. The scintillation fluctuations were highly correlated at the two observatories, but with a varying time lag as seen in Fig. 17.7. A sinusoidal variation of this time lag, corresponding to the rotation of the baseline, and with an amplitude of 18 s, is clearly present, corresponding to a velocity of 370 km s^{-1}. This is the velocity of the pattern moving over the baseline; presumably it is due to a large proper motion of the pulsar (Chapter 8). There are also considerable random fluctuations, which may be due partly to pattern instability.

Fig. 17.7. Time lag between scintillations of PSR 0329+54 recorded at Jodrell Bank and at Penticton. The time lag changes sinusoidally as the position angle P changes with the rotation of the Earth (after Galt & Lyne 1972).

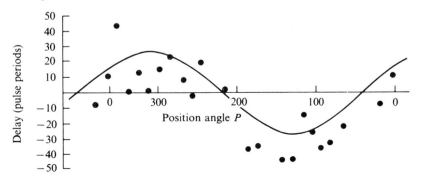

The velocity of the pattern can be deduced from measurements of B_s and τ_s at a single observing station. For a diffracting screen extending along the whole line of sight from a source at distance D

$$B_s = \frac{c}{\pi D \theta_s^2} \text{ and } V_s = \frac{\lambda}{\sqrt{8\pi}\,\theta_s \tau}$$

giving (17.16)

$$V_s = (8\pi c)^{-1/2} (BD)^{1/2} \tau.$$

If we can regard the diffracting irregularities as stationary and unchanging, V_s measures the transverse velocity of the source, combined with a component of the Earth's orbital velocity, which is usually comparatively small. A comparison has been carried out between values of V_s and the measured transverse velocities of pulsars obtained from their proper motions (Lyne and Smith 1982). This shows (Fig. 17.8) a good correlation, and it is now acceptable to use values of V_s as a measure of pulsar velocity.

For most pulsars the proper motion is much larger than the Earth's orbital velocity, but for some there should be a considerable variation of fading speed through the year as the vector addition of the velocities varies; for example, the fading speed for PSR 1929+10 should vary by a factor of about ten through the year. A consistent set of observations to demonstrate this effect has yet to be made.

The corresponding effect in which fading speed varies as a pulsar moves in a binary orbit has been used by Lyne to measure the elements of the orbit (Chapter 10).

Fig. 17.8. Correlation between transverse velocities of pulsars, deduced from their scintillation, and the transverse velocities obtained from their proper motions.

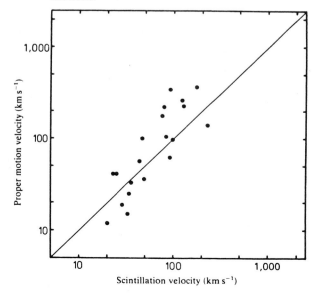

17.9 Pulse broadening

Scintillation is the most obvious, but not the only effect of propagation through irregularities in the ionised interstellar gas. Observations of distant pulsars, and particularly at low radio frequencies, which may be expected to show extreme effects of scintillation, also show the related phenomenon of pulse broadening. Fig. 17.9 shows an example, in which a pulse is drawn out into a long tail at low frequencies. A further example is provided by the Crab Pulsar (Chapter 7), which at frequencies below about 50 MHz shows as a continuous source, in which the pulses are almost completely smeared out by this lengthening process.

The close relation between scintillation and pulse broadening is shown by the rough equality

$$2\pi B_s \Delta \tau \approx 1 \tag{17.17}$$

where B_s is the bandwidth of the scintillation, and $\Delta\tau$ is the pulse broadening. This relation, discussed by Lang (1971), is proved experimentally for several pulsars, although in practice it can only be checked by using measurements of B_s and $\Delta\tau$ obtained at widely different frequencies, with an extrapolation of their known frequency dependence to bring them to the same frequency. The relation is easily understood: if ray paths reach the observer over distances differing by $\Delta\tau$, then the relative phases of the various signal components will range over $(2\pi/\lambda)c\Delta\tau = 2\pi\nu\Delta\tau$, so that the relative phases of the interfering waves will change by the order of 1 radian over a band $B_s \approx 1/2\pi\Delta\tau$.

Fig. 17.9. Pulse lengthening by scattering in the interstellar medium. The pulse profile of PSR 1946+35 at 610 MHz and at 240 MHz.

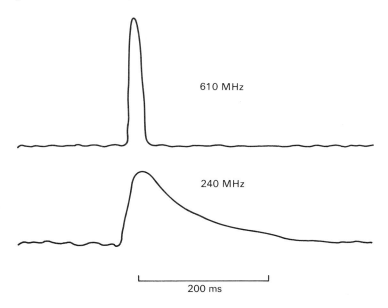

Theoretical treatments of pulse broadening take either the geometrical or diffraction approaches to the problem. A geometric optics approach, involving ray theory, is appropriate for the multiple scattered and greatly broadened pulses, while the diffraction approach may be needed for cases in which multiple scattering is not fully established. We shall consider treatments of these by Williamson (1973) and Uscinski (1974).

17.10 Multiple scattering

Consider first a thin slab of scattering material approximately halfway between the source and the observer, separated by distance L (Fig. 17.10). If the slab behaves as a thick scatterer, in the sense that it introduces large random phase changes, then emergent rays are scattered with a Gaussian angular distribution, with probability

$$P(\theta) \propto \exp(\theta/\theta_0)^2. \tag{17.18}$$

The delay t introduced into a ray deviated by angle θ is $(L/4c)\theta^2$. The probability of a ray deviated between θ and $\theta+d\theta$ reaching the observer is proportional to $\theta \, d\theta$, because of the solid angle subtended by the screen. The probability of a delay $P(t)$ is therefore given by

$$P(t)dt \propto \theta \exp\left(\frac{\theta}{\theta_o}\right)^2 d\theta. \tag{17.19}$$

Hence the distribution of delay is

$$P(t) \propto \exp\left(\frac{t}{\tau}\right). \tag{17.20}$$

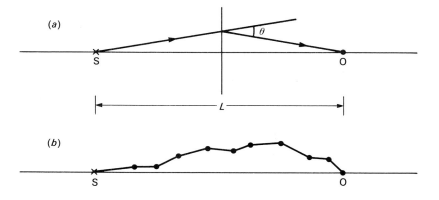

Fig. 17.10. Geometry of multiple scattering: (*a*) single scattering; (*b*) multiple scattering. S, source; O, observer.

17.11 Observations of pulse broadening

A pulse will therefore be observed to have a sharp rise and an exponential decay, with time scale $\tau = L\theta_0^2/4c$. The time scale is therefore proportional to v^{-4} for a screen with a single scale size a. It is interesting to note that the apparent size of the source changes during the pulse; the sharp leading edge comes from a central point, while the exponential trail comes from an expanding halo of progressively more scattered rays.

The effect is similar for a screen over a range of positions roughly midway between the source and observer, but if it is close to either source or observer the time scale is shortened proportional to the shorter of the two distances. Extension of the analysis to a scattering medium filling the whole space involves a more complicated ray tracing problem (Fig. 17.10(b)). Williamson (1973) obtained the delay distributions in Fig. 17.11 from a random walk analysis, and Uscinski (1974) developed the diffraction analysis to produce similar although not identical results.

17.11 Observations of pulse broadening

The Vela Pulsar PSR 0833−45, has been observed by Ables, Komesaroff and Hamilton (1973) at six frequencies ranging from 150 to

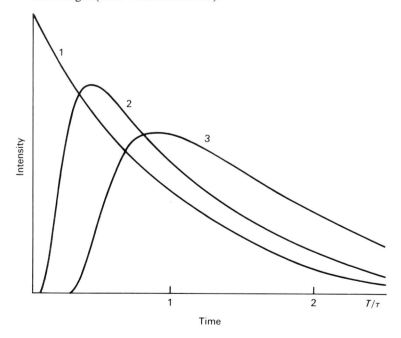

Fig. 17.11. Pulse broadening functions. 1, Scattering from a thin slab; 2, scattering from a more extended region; 3, scattering from irregularities filling the whole line of sight (after Williamson 1973).

1420 MHz. The pulse shapes (Fig. 17.12) show pulse lengthening with a time scale $\tau_0 = 2.2 \times \nu^{-4}$ ms (ν in MHz). The variation with frequency is as expected. A detailed examination of the pulse shapes by Williamson (1974) provides an interesting comparison with the scattering theory. Fig. 17.13 shows the observed pulse shape at a low frequency (297 MHz) superposed on a theoretical shape derived by convolving the unscattered pulse shape, as seen at high frequencies, with the scattering function for an extended scattering medium localised about either the source of the observer.

Although, as expected, it is difficult to distinguish between the three cases of Fig. 17.11 for this pulsar the third does not provide a fit to observation, and it seems that along the line of sight to this pulsar the irregularities cannot be distributed uniformly.

A clearer indication of the same effect is obtained from pulse broadening in the Crab Pulsar. Here Sutton, Staelin and Price (1971) found that individual strong pulses, which are probably less than 1 ms long at the source, show a very sharp rise followed by an exponential decay with time constant $\tau_0 = 12.2$ ms at 115 MHz. The shape of this broadening is certainly

Fig. 17.12. Pulse lengthening in the Vela Pulsar. The traces cover the full period of 89 ms (after Ables *et al.* 1973).

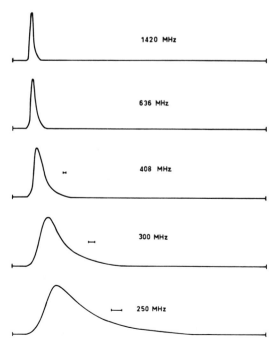

17.11 Observations of pulse broadening

not consistent with the uniform thick scatterer; it indicates instead that the scatterer is concentrated into a small fraction of the line of sight.

In 1974 the pulse broadening of the Crab Pulsar increased dramatically during a period of only a few weeks (Lyne and Thorne 1975). At 408 MHz the broadening pulse length increased to 4 ms (Fig. 17.14); normally the broadening only lengthens the pulse by 50 μs. Again the shape showed a sharp rise followed by an exponential decay, and again the interpretation

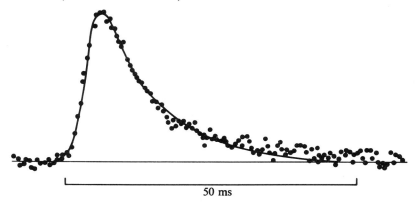

Fig. 17.13. The broadened pulse of the Vela Pulsar at 297 MHz compared with theory for scattering in a region localised either around the source or around the observer (after Williamson 1974).

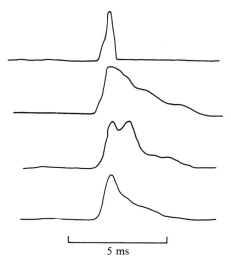

Fig. 17.14. Variable scattering in the Crab Nebula. Recordings at 408 MHz of the interpulse of the Crab Pulsar over a period of 6 weeks (after Lyne & Thorne 1975).

must be that the scattering occurs in a concentrated region. The lengthened profile contained several discrete components, indicating that the scattering region contained a simple structure; it was almost certainly within the Crab Nebula itself. The movement of the region across the line of sight may be part of the expansion of the nebula, or it may be associated with the active 'wisps' near the pulsar (Chapter 1).

Another indication that the scattering irregularities are not distributed uniformly is the difference between broadening observed in different pulsars with similar dispersion measures. Lyne (1971) pointed out for example that PSR 2002+31, with DM = 235, shows no detectable broadening at 408 MHz, while PSR 1946+35 with DM = 129, shows a large effect.

Since the irregularities are evidently more or less localised, it might be postulated that they are close to the pulsar. This may be tested by measuring the angular diameters of the pulsars which show pulse lengthening. We now discuss the relation between scintillation and apparent angular diameter, both in theory and observation.

17.12 Apparent source diameters

The scattering in the interstellar medium, which is responsible for scintillation and pulse lengthening, also increases the apparent angular size of the source, as in the familiar examples of optical 'seeing' in telescopes and the radio diffraction in the solar corona. For the simple example of a physically thin screen halfway between the source and the observer (Fig. 17.1), the source appears on average as a Gaussian disc with half-width θ_s. If the same screen is moved from the half-way position, the apparent angular size θ_s' varies according to the distance from the source, as in Fig. 17.15, where $\theta_s' = \theta_s D/2(D-d)$.

For multiple scattering in a medium filling the whole path Williamson

Fig. 17.15. The apparent size θ_s' of a scattered source S, depending on the distance d of the screen in relation to the observer's distance D.

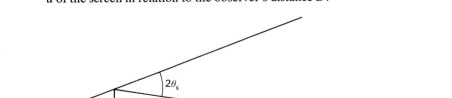

(1974) showed that the mean square image diameter of θ_s^2 is $\frac{1}{3}\phi^2 D$, where ϕ^2 is the mean square scattering angle per unit path length.

The angular size θ_s is small: for the Crab Pulsar it is of the order of 100 milliarcseconds at 100 MHz, varying approximately as the square of the wavelength. The angular diameter of the pulsar itself is, of course, many orders of magnitude smaller at about 100 picoarcseconds. The scattered diameter can be resolved by very long baseline interferometry (VLBI). Vandenberg *et al.* (1973) measured a diameter of $0\rlap{.}''07 \pm 0\rlap{.}''01$ at 115 MHz, and Mutel *et al.* (1974) measured a diameter of $1\rlap{.}''30^{+0.23}_{-0.13}$ at 26 MHz. Unfortunately both these measurements involved only a single baseline, and in neither case was the length of the baseline well suited to the angular diameter. A detailed measurement of the shape of a scattering disc has only been made for Cyg X–3, a radio source in the Galaxy that occasionally flares with very high intensity (Wilkinson *et al.* 1988).

Interstellar scattering has been observed for extragalactic radio sources at 81 MHz by Duffett-Smith and Readhead (1976), continuing the pioneering work by Hewish, which led to the discovery of pulsars. They found that no extragalactic source had an apparent diameter less than $0\rlap{.}''15 \pm 0\rlap{.}''05$. This result is reasonably consistent with the interferometer measurements. Finally, the observed pulse lengthening already discussed in Section 17.11 is simply related to θ_0. For a uniformly filled path, the pulse lengthening τ is given by

$$\tau = \frac{L\theta_s^2}{4c} \tag{17.21}$$

For the Crab Pulsar at 111.5 MHz $\tau = 13$ ms, giving $\theta_s = 0\rlap{.}''1$, in good agreement with the interferometer measurements. The agreement is, in fact, good enough to exclude any configuration in which the scattering is concentrated mainly near the pulsar. The variable component of lengthening observed for the Crab Pulsar therefore must have a different origin, since the speed of variability means that it is concentrated in a small volume, presumably within the Nebula itself.

17.13 Long-term intensity variations

Soon after pulsars were discovered it became evident that their intensities varied over much longer time scales than those involved in scintillation. These long term variations could only be studied by a long series of observations, spanning many months, and their investigation was left to one side while the observers were occupied with problems with more obvious solutions. The origin of long term intensity variations was, meanwhile, attributed loosely to the radiation mechanism in the pulsar itself, or possibly to propagation conditions close to it.

The first clue to the origin came from Sieber (1982), who pointed out a close correlation between the time scale of the intensity fluctuations and the dispersion measures of a set of pulsars previously observed by Helfand. This could only mean that the fluctuations were generated in the interstellar medium, and must be regarded as another form of scintillation. The long time scale showed, however, that very much larger linear scales of electron density fluctuations must be involved. Sieber also pointed out that the time scale seemed to decrease with increasing frequency, which was contrary to previous experience and to the theory of conventional scintillation.

The explanation came from Rickett, Coles and Bourgois (1984), who pointed out that propagation of laser light through turbulent air had already shown a comparable effect, in which there were two coexisting regimes of angular scattering. They were then able to relate the dual nature of radio scintillation to the known existence of a wide continuous spectrum of irregularities, following a power law close to the Kolmogorov law (Section 17.5). The slow variations are now known as refractive scintillation.

We noted in Section 17.7 that frequency drifting in the dynamic spectra of pulsars is related to refraction in large scale irregularities, coexisting with scattering in small scales. The two scale sizes are immediately to the right of the two limit lines in Fig. 17.3; these correspond respectively to the limit of deep phase modulation and to the limit of multipath propagation. Significantly, the second of these varies as $(\lambda D)^{1/2}$, which is reasonably close to the observed dependence of the long term time scale on wavelength and dispersion measure.

Rickett *et al.* were then able to attribute the long term intensity variations to a focussing effect in electron clouds at about the Fresnel distance (Section 17.4). The form of the variations is presumably Gaussian, although focussing can produce occasional unexpectedly high peaks of intensity. The remarkable feature is that for a given pulsar the two time scales t_s, for conventional scintillation and t_c for refractive scintillation are related simply by

$$t_s t_c = \text{constant}.$$

Observers are accustomed to t_s becoming rapidly shorter at longer wavelengths; the surprise is that t_c becomes longer at longer wavelengths. Furthermore, there is an obvious application to other radio sources, provided that these have small enough angular diameters. There are long term

17.13 Long-term intensity variations

variations of intensity in certain small diameter extragalactic sources, known as the low frequency variables. Provided that they have components with angular diameters less than a few milliarcseconds, their intensities will be modulated by focussing in the interstellar medium in our Galaxy, with a time scale close to that observed. Another possibility is that a small source in the centre of a galaxy is modulated by the interstellar medium in its own surroundings.

18

The interstellar magnetic field

The magnetic fields which are first brought to mind in any discussion of pulsars must surely be the enormous fields at the surface of the neutron star itself. These fields are of the order of 10^{12} gauss (10^4 tesla) for most pulsars, and 10^8 gauss for the older, millisecond pulsars. Pulsar radio waves may, however, be used as a means of measuring the magnetic field on the line of sight to the observer; this interstellar field is, by way of extreme contrast, of order 1 to 10 microgauss. The observational link between these extremes is easily stated: the high field of the pulsar is responsible for the linear polarisation of the radio emission, while the Faraday rotation of the plane of polarisation in interstellar space is a measure of the interstellar magnetic field along the line of sight.

18.1 Optical and radio observations

The existence of an interstellar magnetic field was first demonstrated by the observation of the polarisation of starlight, notably by Hall and Mikesell (1950) and by Hiltner in 1949. A small degree of linear polarisation was found, which was common to several stars in the same general direction. The polarisation was due to scattering in interstellar space by particles which are elongated, so that they scatter anisotropically, and which are aligned with their long axes perpendicular to the galactic magnetic field. This alignment occurs through an interaction between the particles and the magnetic field: any component of spin which brings the long axis periodically in and out of alignment with the field will decay. The plane of polarisation therefore provides a very useful measurement of the orientation of the field, but not of its strength.

Radio emission from the Galaxy was already known at the time of discovery of optical polarisation, but it was not until the synchrotron theory of its origin was developed in the Soviet Union a few years later that the existence of radio emission was used as evidence for a general magnetic

18.1 Optical and radio observations

field. The field strength was then estimated to be about 10 microgauss, although this value was derived on the basis of admittedly inadequate information on the flux of cosmic ray electrons. The most significant outcome of the theory was the realisation that the background radiation should be partially linearly polarised. The plane of polarisation should be related to the direction of the field, and the degree of polarisation should be related to the degree of organisation of the field.

Successful measurements of polarisation of the radio background radiation have been made over a range of wavelengths from 21 cm to 2 m. At the longer wavelengths there are severe effects from Faraday rotation between the source and the observer. Faraday rotation is best studied directly by the pulsar observations which are the main subject of this chapter, and we shall not therefore be concerned here with the long wavelength background polarisation. At short wavelengths, where Faraday rotation is small, the degree of polarisation exceeds 10% in several regions of the sky, indicating a well aligned field over a considerable distance. The strongest polarisation is observed at galactic longitude $l = 140°$, where the line of sight appears to be perpendicular to the field over a distance of several hundred parsecs.

A very direct measurement of the magnetic field in some localised regions is provided by the Zeeman effect. The 21 cm radio spectral line of neutral hydrogen is split in a magnetic field, giving two lines of opposite hands of circular polarisation separated by 2.8 Hz per microgauss. This has been observed in absorption, when the widths of some individual absorption components are small enough to allow the split to be observed. The field strengths measured in this way are often greater than 10 microgauss: they refer, of course, to individual absorbing clouds where the density is locally high, and the field strength is not typical of the interstellar medium.

The other method of measuring the interstellar magnetic field was available before the discovery of pulsars. Faraday rotation of the plane of polarisation of extragalactic radio sources can be observed for sources over the whole celestial sphere. The rotation is given by

$$\theta = R\lambda^2 \text{ rad}$$

where R is the rotation measure of a particular source in units of rad m^{-2} and λ is the wavelength of observation. R is determined by the electron density n_e and the component of field along the line of sight, so that

$$R = 0.81 \int B\, n_e \cos\theta\, dL \qquad (18.1)$$

where B is measured in microgauss, n_e in cm^{-3}, and L in parsecs.

The most serious problem in interpreting the observations of Faraday rotation in extragalactic sources is in isolating the galactic component since there are large rotations intrinsic to some of the sources themselves. A clear pattern of variations of R over the sky does, however, emerge from surveys of a sufficient number of sources. Gardner, Morris and Whiteoak (1969) obtained from this pattern a field directed along the local spiral arm, approximately towards longitude $l = 80°$. A later review by Simard-Normandin and Kronberg (1980) of a larger body of data shows that this pattern is overlaid by several complicating features.

If the field were parallel to $l = 80°$ at all distances, the pattern of R over the sky would be very simple, with a clear division of sign between two hemispheres. The intrinsic rotations of the sources would obscure this pattern, but there are obviously also some large anomalies which indicate large scale irregularities in the field pattern. The strength of the field was not obtainable directly from these Faraday rotation measurements, since the electron density n_e remained unknown.

An account of the various pieces of evidence concerning the interstellar magnetic field, as they stood in 1968 at the time of the discovery of the pulsars, was given by van der Hulst (1967).

18.2 Faraday rotation in pulsars

The discovery of a high degree of linear polarisation in pulsar radio signals (Lyne and Smith 1968) led at once to the most direct way of measuring the interstellar magnetic field. Faraday rotation of the plane of polarisation can be measured, as for the extragalactic radio sources, by measuring the position angle of the polarisation at several radio frequencies. In contrast to the extragalactic radio sources, there is no rotation intrinsic to the pulsar, so that the uncertainty in the origin of the rotation is removed. The main advantage over the extragalactic sources is, however, the availability of the dispersion measure DM at the same time as the rotation measure R. This removes most of the uncertainty in interpreting equation 18.1, since it provides a value for the electron content in the line of sight.

The ratio R/DM, where

$$R = 0.81 \int n_e B \cos\theta dL, \quad DM = \int n_e dL. \tag{18.2}$$

gives a value of field $\langle B \cos\theta \rangle$, which is an average along the line of sight. The average is weighted by the electron density along the line; if, for instance, a large part of the dispersion measure in a particular pulsar is due to a single HII region, then the field in that region receives a heavy weighting. If a pulsar lies at a large distance above the galactic plane where the

18.3 The configuration of the local field

electron density may be small, its rotation measure does not depend much on the magnetic field in its immediate vicinity, but mostly on the field within one scale height of the electron density distribution. The method is obviously difficult to apply if the strength or direction of the field vary considerably along the line of sight; fortunately the main structure of the field is on a large scale, and the average field $\langle B \cos\theta \rangle$ turns out to be a most useful measure.

Faraday rotation is measured from the variation of polarisation position angle with radio frequency. This is usually obtained over a wide frequency range, although at low frequencies where the rotation is large it may be convenient to measure the small change in position angle $\delta\theta$ between adjacent frequencies v, $v+\Delta v$. If v is in MHz

$$\Delta\theta = -2R(300/v)^2 . \Delta v/v.$$

At frequencies near 300 MHz a frequency separation Δv of a few megahertz is appropriate for most rotation measures; the more distant pulsars with large rotation measures may require observations about 1 GHz.

The first measurements of pulsar Faraday rotation (Smith 1968) referred to PSR 0950+08, which had recently been found to be highly polarised. Unfortunately the line of sight to PSR 0950+08 is nearly transverse to the local magnetic field, and this pulsar has almost the smallest known rotation measure; it is so small that the measured rotation could be accounted for by ionospheric rotation alone. Lyne, Smith and Graham (1971) subsequently obtained line of sight field values up to 3 microgauss for several pulsars. Data for 160 pulsars are now available (Hamilton and Lyne, 1987)

18.3 The configuration of the local field

A reasonably uniform set of data covering the whole sky has been presented and analysed by Manchester (1974). Most of the 40 pulsars in this set are comparatively local, as can be seen from their small dispersion measures. The magnetic field components obtained from these local pulsars are shown in Fig. 18.1. These components are shown as circles whose area is proportional to field strength. The circles containing a cross represent positive rotation measure, corresponding to fields directed towards the observer.

There is a remarkably clear-cut division between positive rotation measures, which occur in the 180°–360° hemisphere, and negative rotation measures, which occur in the 0°–180° hemisphere. This indicates that the field is well aligned out to a distance of several hundred parsecs, and that it is directed tangentially, roughly towards $l = 90°$. There are some irregularities, as might be expected from work on the polarisation of background radio emission; Wilkinson and Smith (1974), Spoelstra (1984), and Beuermann et al. (1985) show that irregularities exist on a scale of some tens of parsecs, and with a strength comparable with the mean field. There is probably also an extensive irregularity associated with the North Galactic Spur, which is believed to be the remnant of a supernova shell. By omitting pulsars at high galactic latitudes and in the North Galactic Spur, Manchester obtained a local field strength of 2.2 ± 0.4 microgauss, directed towards longitude $l = 94°±11°$.

This average field refers to distances up to a few hundred parsecs. Even within this distance there are considerable deviations in field direction; the pattern of background polarisation shows very clearly that a line of sight towards $l = 140°$ somewhere within 1 kpc intersects the field at right-angles, indicating a field direction towards $l = 50°$ (Bingham and Shakeshaft, 1967).

The more distant field is not yet well determined: we include the observational work which is still needed on high dispersion pulsars in our résumé of promising new pulsar research in Chapter 19. Preliminary indications are that a complete reversal in field direction can be delineated in the inner regions of the Galaxy.

Fig. 18.1. The interstellar magnetic field. The field is the weighted mean along the line of sight to pulsars whose Faraday rotation has been measured (Manchester 1974). The field strength is indicated by the size of the circles; the circles with a cross indicate a field directed towards the observer.

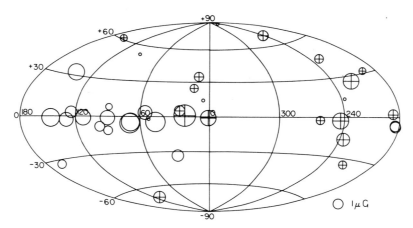

19

Achievements and prospects

The first two decades of pulsar research have seen considerable achievements. Since Jocelyn Bell saw the first record of radio pulses as a mysterious 'piece of scruff' on a chart recording, bringing the neutron star from a remote world of theory into reality, we have been able to investigate the structure of the insides and the outsides of neutron stars in astonishing detail. We now have proof of their origin in supernova explosions, and we have good evidence of their evolution in binary star systems. The combination of pulsars and binary X-ray sources has opened a whole new branch of astrophysics concerned with accretion onto condensed stars. Pulsars have provided an entirely new means of investigating the interstellar medium, measuring the total electron content through dispersion, irregularities in electron density through scintillation, and the magnetic field through Faraday rotation. The millisecond pulsars have provided clocks of unparalleled regularity, enabling the most sensitive tests of relativity theory to be made. Nevertheless the field of pulsar research is as active as ever, and major questions remain to be answered.

In this chapter we set out some of the current issues, and describe how they are being approached.

19.1 The population and the birthrate

As described in Chapter 10, the birth, evolution, and extinction of pulsars may be investigated by observing their distribution in period P and the rate of slowdown $\dot{P}$. It is reasonable to assume that the population is statistically stationary, and that the periods of most pulsars are increasing monotonically. If the further assumption is made that all pulsars start life with a very short period, say 10 milliseconds or less, there will be a flow of pulsars through the observed distribution in period such that the rate of crossing all periods is the same, and equal to the birthrate. Obviously the observed flow rate is lower at long periods, due to the death of pulsars (or to a reduction in their intensity, which may make them difficult to observe). The statistics at long periods can therefore be used to investigate the decay of the magnetic field and the cessation of radio emission.

The difficulty in these population statistics comes for the short-period pulsars, whose periods are increasing rapidly. Apart from the millisecond pulsars, very few are known with periods less than 100 milliseconds. They are of course comparatively rare because they slow down rapidly and spend only a short time with such periods, but their population is particularly important. Some authors have suggested that there are fewer of these short-period pulsars than would be expected from the flow rate at longer periods, which would imply that some pulsars are born with longer periods and are 'injected' into the flow (Vivekenand and Narayan 1981). This should not affect the birthrate derived from the flow at longer periods, provided the injection regime does not overlap with the region where the emission is fading, but it is at least an interesting question regarding the dynamics of the stellar collapse and rotational speedup.

The millisecond pulsars are evidently not part of the normal population. They are regarded as rejuvenated pulsars whose age is demonstrated by the weakness of their magnetic fields; whether or not they had traversed the normal flow of pulsars, the theory that they were spun up to their present high rotation rates through the accretion of material from a binary companion receives obvious support from the existence of the binary pulsars, and from the relation between the globular cluster pulsars and the low-mass X-ray binaries. The solitary millisecond pulsars may also have been spun up in binary systems and subsequently lost their companions either through explosion or by evaporation, a process which is occurring in the eclipsing binary PSR 1957+20. The evaporation process may also account for the lack of binaries in the main pulsar population (van den Heuvel and van Paradijs 1988).

An alternative theory of the origin of solitary pulsars, whether in the main population or among the millisecond pulsars, is through coalescence of a binary system consisting of a white dwarf and a neutron star (van den Heuvel and Bonsema 1984). These various routes to solitary pulsars are the subject of much current theoretical work.

The challenge for the observers is evidently to improve the statistical sample of short-period pulsars, so as to establish the flow of truly young pulsars into the normal population and the routes by which millisecond pulsars can be created. It is, of course, the short-period pulsars that are most difficult to discover; the difficulty is compounded by the need to search towards the Galactic centre.

19.2 Pulsar searches

Most of the known pulsars were discovered in surveys at comparatively low frequencies, usually near 400 MHz, where pulse broadening due to interstellar scattering is often very large for all but the fairly local pulsars. These surveys were therefore insensitive to the distant, short-period pulsars, and especially those lying in the galactic plane, where the sensitivity of the searches was also reduced by the high level of galactic background radiation. An example of a recent survey designed to overcome this problem, mainly by the use of the higher radio frequency of 1400 MHz, is the Clifton and Lyne survey described in Chapter 3. A later survey, in progress at the Parkes radiotelescope in Australia as this book is being written, combines the use of this higher radio frequency with a large bandwidth to obtain greater sensitivity.

Although the use of a higher frequency reduces the problem of pulse broadening, it must contend with the low signal strengths due to the steep spectrum of the radio emission. The Parkes survey therefore uses the wide signal bandwidth of 300 MHz, divided into a number of separate channels, as described in Chapter 3, to allow for a range of dispersion measures. Furthermore, the signal sampling rate must be very rapid to allow the detection of short period pulsars, and the integration time of a single observation must be long so that weak pulsars can be detected. This survey therefore uses 80 adjacent frequency channels, sampled at intervals of 0.3 milliseconds, with an integration time of 200 seconds. More than 50 million samples are required for the observation of a single beam area, i.e. about 0.05 square degrees of sky. The massive computation required in searching for periodic signals in this mass of data, including a range of dispersion measures, demands the use of one of the very large supercomputers that fortunately are now available. The survey itself is nevertheless slow, since the telescope beam is so narrow.

Organised searches of this kind have not so far led to dramatic changes in the population statistics presented in Chapter 8; significant improvements to the catalogues of pulsars, including the millisecond pulsars, may most likely arise from extensions of the southern hemisphere surveys using the 210-ft Parkes telescope.

19.3 Condensed matter within pulsars

The theory of the interior of neutron star has been generally well confirmed by the observations of the rotational dynamics of pulsars. There are, however, some interesting details still to be resolved, and we should ask what are the observations that could be most helpful. On a long timescale there are the questions of magnetic field decay and dipole alignment; the theory of these has been discussed by Jones (1987). On a shorter timescale there are the glitches and oscillations described in Chapter 6; these are our closest equivalent to the classic approach of observing the response of a system to a perturbation. What can we hope to provide in further observations of these phenomena?

The simple answer is that there is no substitute for an extended and consistent series of timing observations. Fortunately we already have several detailed records of glitches in the Vela Pulsar; on the basis of these the theory of vortex drift and pinning has been built. Glitches are much less common in most pulsars, but the few examples suggest that there is still much to understand. In the Crab Pulsar glitches apparently occur at intervals of around five years; but only one resulted in a long-lasting change in slowdown rate. Two successive glitches occurred only two years apart in the much older PSR 0355+64; the second was one of the largest seen in any pulsar. The oscillations with a period of about 20 months in the rotational phase of the Crab Pulsar may have opened a new method for the investigation of the superfluid interior, in which the vortex lattice may be oscillating in the same way that can be observed in laboratory experiments in liquid helium; are these oscillations real, and can they be observed in other pulsars with longer periods?

It is to be hoped that the long-term observation programmes that are needed for the resolution of these questions will remain part of the programmes of the radio observatories for at least another decade.

19.4 Young pulsars and supernova remnants

Although the discovery of the Crab and Vela Pulsars proved conclusively that supernova explosions involved the collapse of a stellar core to form a neutron star, it is by no means clear that all pulsars originate in explosive collapse of this kind. The rate of occurrence of supernovae in the Galaxy is of order one per 100 years, which is about equal to the birthrate of pulsars, as derived in Chapter 10. Neither of these rates, however, is known to better than a factor of two, and it would be wrong to assume that we can equate the two. Furthermore, it now seems that there may be a variety of types of stellar collapse that can result in the formation of a neutron star. The classical picture of a star with mass greater than about 8 $M_\odot$, evolving through the red giant phase as described in Chapter 9, provides an inadequate explanation for most of the supernovae which have been studied in detail; for example, SN 1987A in the Large Magellanic Cloud appears to have originated in a blue giant, while there is a strong

19.4 Young pulsars and supernova remnants

suspicion that the lack of visual observations of the supernova Cas A means that it was very under-luminous. Theories are now being developed of 'quiet collapse', or 'silent supernovae', resulting from the accretion of material on to a white dwarf star in a binary system. Again, as we have mentioned as a possible origin of solitary millisecond pulsars, there is a theory that a pair of white dwarfs in a close binary system might lose orbital energy by gravitational radiation, coalesce and collapse into a neutron star. How is the relative importance of these possibilities to be tested?

Only a small number of pulsars have been found in observable supernova remnants. There are, however, at least a hundred supernova remnants in which we might hope to find a pulsar. The three most recent in the Galaxy, Tycho's, Kepler's, and Cas A, have been searched for radio point sources which might be pulsars. None has been found. The radio source 3C 58 is a supernova remnant like the Crab, but no pulsar has been found in it. Some pulsars will, of course, be aligned so that their radio radiation is beamed away from an observer in the Solar System; as we have seen in Chapter 15, optical radiation from the young pulsars is less tightly beamed, and possibly an optical search may yield a pulsar that is concealed from radio searches.

This imperfect match between supernova remnants and pulsars leads to searches which start both in the radio sky maps and in the optical spectrum. The example is CTB 80, a radio nebula containing a point-like radio source which subsequently was found to be a pulsar. We need more such associations. Optically, a further search for pulsating stars in the recent supernovae may become worthwhile when the new generation of very large telescopes comes into operation. Meanwhile there is an intensive search for a pulsar in the remains of the recent supernova SN1987A. Not only is this predicted by the general theory of stellar collapse in the supernova explosion, there is already also some evidence that there is a continuing source of energy in the remnant, in addition to the decaying energy from the explosion itself. A pulsar within the remnant would be hidden from radio observation by the expanding cloud of ionised gas, and might only be observable after some tens of years. Optical observations offer the best opportunity, and the first report of the pulsar came from observations in January 1989 with the 4-m optical telescope at Cerro Tololo (Middleditch *et al.* 1989).

The pulsar in SN1987A is reported to have a period of 0.508 milliseconds. It is therefore spinning about three times faster than any other pulsar, so fast indeed that it could only be dynamically stable if it were very condensed, i.e. that its material had a 'soft' equation of state (see Chapter

2). The rapid rotation also calls into question the evolutionary sequence outlined in Chapter 5, since the accepted view is that, after two years of slowdown, the period should have lengthened to several milliseconds. Confirmation of the observations will evidently lead to a new series of papers explaining the rapid rotation and slow spindown of this new pulsar.

19.5 The interstellar medium

We have already recounted in Chapter 17 how pulsar observations give us the mean electron density and the magnetic field in the local interstellar medium, and how scintillation observations are related to the turbulent structure of the ionised clouds. Problems of great astrophysical interest remain in all these areas. We already need a better model of the electron density distribution, extending further from the galactic plane and closer to the galactic centre; this can be achieved through the identification of pulsars with objects such as the globular clusters whose distances are known (Chapter 3). The magnetic field could also be traced to much greater distances, beyond the local region where it is aligned with the local spiral arm structure. There is already a sufficient body of data from Faraday rotation measurements of the more distant pulsars to show that the field direction reverses as the line of sight extends into the inner spiral arms, in directions between longitudes 0° and 60° (Lyne and Smith 1989). More observations of pulsars from the southern hemisphere are needed for the interpretation of this reversal, although it is already clear that it occurs on a large scale.

The distribution of electron irregularities is usually described in terms of a power law relating the density to the scale size. At present the observations fit reasonably well to the spectrum expected from Kolmogorov turbulence, in which the turbulence is generated at large scales and degenerates to progressively smaller sizes by simple mechanical processes. The origin of the turbulence is presumably the shock waves from supernovae. This simple picture may be tested in several ways.

At the larger scales of turbulence, corresponding to the longer time scales of fluctuation, there are already reports (Fiedler *et al.* 1987) of slow fluctuations in the intensities of some quasars which indicate the existence of discrete, non-random electron clouds. The long-term monitoring observations of quasars and pulsars which address this question are inevitably tedious, but it seems possible that they may reveal the injection process more clearly than studies of the small-scale random turbulence into which these larger clouds decay.

The detail of the smaller scale irregularities may nevertheless be worth

19.6 The emission mechanism

further study. The Kolmogorov spectrum itself is not well established. It is also possible that the irregularities are not isotropic; here an interesting development may come from the observations of scintillation in binary pulsars, where the direction of the pulsar velocity changes round the binary orbit. Most of the resultant cyclic change in scintillation rate is attributable to the changing magnitude of the transverse velocity, but it already appears that the scintillation rate for the binary pulsar PSR 0655+54 also varies with the direction of motion as expected for anisotropic turbulence in the interstellar scattering.

19.6 The emission mechanism

Finally we return to the phenomenon that first revealed the existence of neutron stars and that has provided most of the data on which this book has been based: we refer to the beams of radio emission from the polar caps of the rotating pulsars. We have been able to describe the location of the emitter, and we can relate the radiated energy to the flow of energetic particles from an accelerating gap immediately above the stellar surface. We can describe how the strength of the emission varies, as demonstrated by pulse drifting and nulling, but we do not understand why it varies either in time or in space. What can happen at the accelerating gap which concentrates the radiating particles into the patches which we observe as subpulses? Above all what is the radiation mechanism which produces the astonishingly powerful radio radiation?

Research on neutron stars has made astonishing progress since the first predictions of their existence. Their geometrical properties have been revealed by their rotation, and the irregularities in their slowdown have revealed the physics of their structure. Their evolution has been traced from their birth in a supernova, through the solitary pulsars to the X-ray binaries and the spin-up to become millisecond pulsars. Their total energy output is to be seen typically in the visible radiation of the Crab Nebula. The most detailed of these investigations have come from the radio pulses which gave pulsars their name; it is ironic that it is the rotating lighthouse beam itself that is the least understood of all the observable properties of pulsars.

20

The Pulsar Catalogue

The first pulsars discovered at Cambridge were listed as CP followed by their right ascension, i.e. CP 1919 etc. This practice was followed briefly by other observatories – JP for Jodrell Bank, HP for Harvard, MP for Molonglo, NP for National Radio Astronomy Observatory, AP for Arecibo. The initials PSR, for Pulsating Source of Radio, were first used by Turtle and Vaughan (1968) in announcing the discovery of two pulsars in the southern hemisphere. It soon became accepted that a combined catalogue should adopt the PSR designation throughout. The use of right ascension and declination, as in PSR 0531+21 for the Crab Pulsar, provided a sufficient discriminant for many years; it has however become necessary occasionally to quote declination in tenths of degrees, as in PSR 1913+167. The two pulsars in the globular cluster 47 Tuc are so close together that they are designated PSR 0021−72 A and B.

Manchester and Taylor published complete catalogues in 1975 and 1981, the latter containing 330 pulsars with many observed and derived parameters, together with comprehensive references. Since then the catalogue has been kept up to date by these two authors, joined by Lyne. New parameters are transmitted by electronic mail, and the observing schedules at the major radio observatories have automatic access to the updated information. There are now over 450 entries in the catalogue (March 1989). We are grateful to Manchester and Taylor for permission to compile a condensed version of this joint work.

The catalogue entries in this chapter give the following observed parameters: RA (Right Ascension), DEC (Declination), Galactic Longitude and Latitude, DM (Dispersion Measure), Distance (kpc), Period (seconds), and the logarithms of the Period derivative $\dot{P}$ and Age $\frac{1}{2}P/\dot{P}$ (years). The distance is derived from DM, using a model electron distribution as set out in Chapter 4.

	PSR	RA(1950.0)	DEC(1950.0)	Long (deg)	Lat (deg)	DM	Dist (kpc)	Period (sec)	Pdot	Age	Notes	Association
1	0011+47	00:11:39.8	+47:29:53	116.5	-14.6	30.	1.1	1.240699	-15.3	7.5		
2	0021-72A	00:21:53.0	-72:21:30	305.9	-44.9	65.	2.5	0.004479			B,M	In GC 47 Tuc
3	0021-72B	00:21:53.0	-72:21:30	305.9	-44.9	65.	2.5	0.006127			B,M	In GC 47 Tuc
4	0031-07	00:31:36.5	-07:38:33	110.4	-69.8	11.	0.4	0.942951	-15.4	7.6		
5	0037+56	00:37:41.0	+56:59:57	121.5	-5.6	91.	3.4	1.118224	-14.5	6.8		
6	0042-73	00:42:30.0	-73:30:00	303.6	-43.9	101.	3.8	0.926499				In SMC
7	0045+33	00:45:51.0	+33:55:47	122.3	-28.7	41.	1.6	1.217094	-14.6	6.9		
8	0052+51	00:52:52.2	+51:01:11	123.6	-11.6	43.	1.5	2.115168	-14.0	6.5		
9	0053+47	00:53:33.4	+47:40:02	123.8	-14.9	18.	0.6	0.472036	-14.4	6.3		
10	0059+65	00:59:20.7	+65:21:06	124.1	2.8	67.	2.1	1.679162	-14.2	6.7		
11	0105+65	01:05:05.1	+65:52:32	124.6	3.3	30.	0.9	1.283653	-13.9	6.2		
12	0105+68	01:05:05.8	+68:49:53	124.5	6.3	60.	2.0	1.071118	-16.3	8.5		
13	0114+58	01:14:28.5	+58:58:50	126.3	-3.5	49.	1.5	0.101437	-14.2	5.4		
14	0136+57	01:36:01.7	+57:59:20	129.2	-4.0	74.	2.5	0.272446	-14.0	5.6		
15	0138+59	01:38:17.3	+59:54:22	129.1	-2.1	35.	3.0	1.222948	-15.4	7.7		
16	0144+59	01:44:20.8	+59:07:06	130.1	-2.7	39.	1.1	0.196321	-15.6	7.1		
17	0148-06	01:48:52.5	-06:49:49	160.4	-65.0	25.	1.0	1.464664	-15.4	7.7		
18	0149-16	01:49:46.4	-16:52:39	179.3	-72.5	12.	0.4	0.832741	-14.9	7.0		
19	0153+39	01:53:54.3	+39:34:52	136.4	-21.3	59.	2.3	1.811561	-15.7	8.2		
20	0154+61	01:54:14.9	+61:57:49	130.6	0.3	26.	0.7	2.351653	-12.7	5.3		
21	0203-40	02:03:57.2	-40:42:22	258.6	-69.6	13.	0.5	0.630550	-14.9	6.9		
22	0226+70	02:26:47.1	+70:13:16	131.2	9.2	47.	1.7	1.466819	-14.5	6.9		
23	0254-53	02:54:24.2	-53:16:25	269.9	-55.3	16.	0.6	0.447708	-16.5	8.4		
24	0301+19	03:01:42.4	+19:21:13	161.1	-33.3	16.	0.6	1.387584	-14.9	7.2		
25	0320+39	03:20:10.2	+39:34:16	152.2	-14.3	26.	0.9	3.032072	-15.1	7.8		
26	0329+54	03:29:11.0	+54:24:37	145.0	-1.2	27.	2.3	0.714519	-14.7	6.7		
27	0331+45	03:31:47.6	+45:45:55	150.3	-8.0	46.	1.6	0.269201	-17.1	8.8		
28	0339+53	03:39:25.9	+53:03:22	147.0	-1.4	69.	2.0	1.934475	-13.9	6.4		
29	0353+52	03:53:56.3	+52:28:21	149.1	-0.5	103.	2.9	0.197030	-15.3	6.8		
30	0355+54	03:55:00.4	+54:04:42	148.2	0.3	57.	1.6	0.156381	-14.4	5.8	G	

Notes: B   The pulsar is a member of a binary system.
       M   This is a millisecond pulsar.
       G   This pulsar has displayed glitches in rotation period.

	PSR	RA(1950.0)	DEC(1950.0)	Long (deg)	Lat (deg)	DM	Dist (kpc)	Period (sec)	Pdot	Age	Notes	Association
31	0402+61	04:02:07.9	+61:30:33	144.0	7.0	66.	2.4	0.594571	-14.3	6.2		
32	0403-76	04:03:15.2	-76:16:26	290.3	-35.9	22.	0.7	0.545252	-14.8	6.7		
33	0410+69	04:10:39.2	+69:46:40	138.9	13.7	27.	0.9	0.390715	-16.1	7.9		
34	0447-12	04:47:49.5	-12:53:12	211.1	-32.6	36.	1.4	0.438014	-16.0	7.8		
35	0450+55	04:50:00.2	+55:38:49	152.6	7.5	14.	0.4	0.340729	-14.6	6.4		
36	0450-18	04:50:21.4	-18:04:21	217.1	-34.1	40.	1.6	0.548935	-14.2	6.2		
37	0458+46	04:58:21.9	+46:49:47	160.4	3.1	41.	1.2	0.638564	-14.3	6.3		
38	0523+11	05:23:09.5	+11:12:41	192.7	-13.2	79.	3.2	0.354438	-16.1	7.9		
39	0525+21	05:25:51.7	+21:58:01	183.9	-6.9	51.	2.0	3.745497	-13.4	6.2	G	
40	0529-66	05:29:30.0	-66:57:00	277.0	-32.8	125.	4.9	0.975714				In LMC
41	0531+21	05:31:31.4	+21:58:54	184.6	-5.8	57.	2.0	0.033342	-12.4	3.1	G	Crab SNR
42	0538-75	05:38:18.9	-75:45:38	287.2	-30.8	18.	0.6	1.245855	-15.2	7.5		
43	0540+23	05:40:06.9	+23:27:46	184.4	-3.3	78.	2.6	0.245966	-13.8	5.4		SNR in LMC
44	0540-69	05:45:34.0	-69:21:24	279.7	-31.1	0.	0.0	0.050310	-12.3	3.2		
45	0559-05	05:59:31.6	-05:27:46	212.2	-13.5	81.	3.3	0.395969	-14.9	6.7		
46	0559-57	05:59:59.4	-57:56:52	266.5	-29.3	30.	1.1	2.261365	-14.6	7.1		
47	0609+37	06:09:23.9	+37:22:26	175.5	9.1	27.	0.9	0.297982	-16.2	7.9		
48	0611+22	06:11:14.9	+22:26:06	188.9	2.4	97.	3.3	0.334925	-13.2	4.9		
49	0621-04	06:21:51.4	-04:23:11	213.8	-8.0	72.	2.7	1.039076	-15.1	7.3		
50	0626+24	06:26:02.0	+24:17:57	188.8	6.2	84.	3.2	0.476622	-14.7	6.6		
51	0628-28	06:28:51.8	-28:32:34	237.0	-16.8	34.	1.3	1.244417	-14.1	6.4		
52	0643+80	06:43:53.5	+80:55:31	133.2	26.8	33.	1.2	1.214440	-14.4	6.7		
53	0655+64	06:55:49.4	+64:22:23	151.6	25.2	9.	0.3	0.195671	-18.2	9.7	B	White dwarf
54	0656+14	06:56:57.9	+14:18:37	201.1	8.3	14.	0.4	0.384875	-13.3	5.0		
55	0727-18	07:27:19.4	-18:30:26	233.8	-0.3	61.	1.6	0.510151	-13.7	5.6		
56	0736-40	07:36:50.9	-40:35:45	254.2	-9.2	161.	2.5	0.374919	-14.8	6.6		
57	0740-28	07:40:47.7	-28:15:41	243.8	-2.4	74.	1.5	0.166752	-13.8	5.2		
58	0743-53	07:43:50.5	-53:44:01	266.6	-14.3	122.	2.4	0.214836	-14.6	6.1		
59	0751+32	07:51:29.3	+32:39:52	188.2	26.7	39.	1.5	1.442349	-15.0	7.3		
60	0756-15	07:56:11.3	-15:19:58	234.5	7.2	64.	2.3	0.682264	-14.8	6.8		

Notes: B   The pulsar is a member of a binary system.
       M   This is a millisecond pulsar.
       G   This pulsar has displayed glitches in rotation period.

PSR	RA(1950.0)	DEC(1950.0)	Long (deg)	Lat (deg)	DM	Dist (kpc)	Period (sec)	Pdot	Age	Notes	Association
61 0808-47	08:08:12.6	-47:45:01	263.3	-8.0	228.	5.7	0.547198	-14.5	6.4		
62 0809+74	08:09:02.9	+74:38:13	140.0	31.6	6.	0.2	1.292241	-15.8	8.1		
63 0818-13	08:18:06.0	-13:41:23	235.9	12.6	41.	1.5	1.238128	-14.7	7.0		
64 0818-41	08:18:29.7	-41:05:04	258.7	-2.7	111.	0.7	0.545446	-16.6	8.5		
65 0820+02	08:20:34.0	+02:08:54	222.0	21.2	24.	0.8	0.864873	-16.0	8.1	B	White dwarf
66 0823+26	08:23:50.5	+26:47:19	197.0	31.7	19.	0.4	0.530660	-14.8	6.7		
67 0826-34	08:26:19.1	-34:07:08	254.0	2.6	47.	0.4	1.848918	-15.0	7.5		
68 0833-45	08:33:39.3	-45:00:10	263.6	-2.8	69.	0.5	0.089286	-12.9	4.1	G	Vela SNR
69 0834+06	08:34:26.2	+06:20:44	219.7	26.3	13.	0.4	1.273764	-14.2	6.5		
70 0835-41	08:35:33.3	-41:24:42	260.9	-0.3	148.	2.4	0.751621	-14.5	6.5		
71 0839-53	08:39:09.2	-53:21:52	270.8	-7.1	156.	3.1	0.720612	-14.8	6.8		
72 0840-48	08:40:29.7	-48:40:32	267.2	-4.1	197.	4.0	0.644354	-14.0	6.0		
73 0841+80	08:41:24.8	+80:40:00	132.7	31.5	34.	1.3	1.602228	-15.4	7.8		
74 0844-35	08:44:07.9	-35:22:39	257.0	4.7	92.	0.6	1.116096	-14.8	7.1		
75 0853-33	08:53:38.0	-33:19:58	256.8	7.5	88.	0.6	1.267533	-14.2	6.5		
76 0855-61	08:55:54.0	-61:26:16	278.6	-10.4	95.	3.6	0.962509	-14.8	7.0		
77 0901-63	09:01:31.8	-63:13:17	280.4	-11.1	76.	2.8	0.660313	-16.0	8.0		
78 0903-42	09:03:08.2	-42:34:13	265.1	2.9	146.	1.7	0.965171	-14.7	6.9		
79 0904-74	09:04:28.8	-74:47:41	289.7	-18.3	51.	1.9	0.549553	-15.3	7.3		
80 0905-51	09:05:40.6	-51:45:51	272.2	-3.0	104.	0.9	0.253556	-14.7	6.3		
81 0906-17	09:06:18.8	-17:27:25	246.1	19.8	16.	0.5	0.401625	-15.2	7.0		
82 0906-49	09:06:54.5	-49:00:54	270.3	-1.0	192.	3.2	0.106755	-13.8	5.0		
83 0909-71	09:09:22.3	-71:59:51	287.7	-16.3	54.	2.0	1.362890	-15.5	7.8		
84 0917+63	09:17:15.6	+63:07:00	151.4	40.7	13.	0.5	1.567993	-14.4	6.8		
85 0919+06	09:19:35.1	+06:51:12	225.4	36.4	27.	1.0	0.430614	-13.9	5.7		
86 0922-52	09:22:30.8	-52:49:47	274.7	-1.9	153.	2.7	0.746295	-13.5	5.5		
87 0923-58	09:23:04.9	-58:01:08	278.4	-5.6	60.	1.0	0.739500	-14.3	6.4		
88 0932-52	09:32:46.9	-52:36:03	275.7	-0.7	99.	1.1	1.444771	-14.3	6.7		
89 0940-55	09:40:37.8	-55:39:11	278.6	-2.2	180.	4.9	0.664361	-13.6	5.7		
90 0940+16	09:40:46.0	+16:45:08	216.6	45.4	20.	0.7	1.087418	-15.0	7.3		

Notes: B The pulsar is a member of a binary system.
       M This is a millisecond pulsar.
       G This pulsar has displayed glitches in rotation period.

	PSR	RA(1950.0)	DEC(1950.0)	Long (deg)	Lat (deg)	DM	Dist (kpc)	Period (sec)	Pdot	Age	Notes	Association
91	0941-56	09:41:18.7	-56:43:57	279.3	-3.0	161.	4.9	0.808116	-13.4	5.5		
92	0942-13	09:42:04.4	-13:40:52	249.1	28.8	13.	0.4	0.570264	-16.3	8.3		
93	0943+10	09:43:27.1	+10:05:45	225.4	43.1	15.	0.6	1.097704	-14.5	6.7		
94	0950-38	09:50:11.8	-38:25:02	268.7	12.0	167.	4.0	1.373815	-15.2	7.6		
95	0950+08	09:50:30.5	+08:09:45	228.9	43.7	3.	0.1	0.253065	-15.6	7.2		
96	0953-52	09:53:41.8	-52:50:01	278.3	1.2	157.	3.5	0.862118	-14.5	6.6		
97	0957-47	09:57:30.1	-47:55:23	275.7	5.4	93.	1.3	0.670086	-16.1	8.1		
98	0959-54	09:59:51.6	-54:52:39	280.2	0.1	131.	0.1	1.436568	-13.3	5.6		
99	1001-47	10:01:23.8	-47:32:29	276.0	6.1	98.	1.6	0.307072	-13.7	5.3		
100	1010-23	10:10:10.0	-23:23:02	262.1	26.4	27.	1.0	2.517945	-14.9	7.5		
101	1014-53	10:14:37.4	-53:30:14	281.2	2.5	67.	2.0	0.769584	-14.7	6.8		
102	1015-56	10:15:22.8	-56:06:29	282.7	0.3	439.	13.7	0.503459	-14.5	6.4		
103	1016-16	10:16:15.2	-16:27:07	258.3	32.6	48.	1.9	1.804694	-14.8	7.2		
104	1030-58	10:30:15.0	-58:55:00	285.9	-1.0	419.	15.4	0.464208	-14.5	6.4		
105	1039-19	10:39:12.0	-19:26:07	265.6	33.6	32.	1.2	1.386368	-15.0	7.3		
106	1039-55	10:39:59.4	-55:05:22	285.2	3.0	306.	11.8	1.170859	-14.2	6.4		
107	1044-57	10:44:19.5	-57:58:02	287.1	0.7	240.	7.1	0.369427	-14.9	6.7		
108	1054-62	10:54:28.1	-62:42:45	290.3	-3.0	323.	12.3	0.422446	-14.4	6.3		
109	1055-52	10:55:48.6	-52:10:52	286.0	6.6	30.	0.9	0.197108	-14.2	5.7		
110	1056-78	10:56:27.9	-78:58:20	297.6	-17.6	52.	1.9	1.347402	-14.9	7.2		
111	1056-57	10:56:55.3	-57:26:09	288.3	1.9	108.	3.3	1.184997	-14.4	6.6		
112	1105-59	11:05:51.1	-59:30:49	290.2	0.5	105.	2.7	1.516531	-15.5	7.8		
113	1110-65	11:10:36.3	-65:56:45	293.2	-5.2	249.	9.3	0.334213	-15.1	6.8		
114	1110-69	11:10:54.1	-69:10:11	294.4	-8.2	148.	5.5	0.820484	-14.5	6.7		
115	1112+50	11:12:48.4	+50:46:36	154.4	60.4	9.	0.3	1.656438	-14.6	7.0		
116	1114-41	11:14:20.6	-41:06:21	284.5	18.1	41.	1.5	0.943155	-14.1	6.3		
117	1118-79	11:18:11.1	-79:20:04	298.7	-17.5	27.	0.9	2.280597	-14.4	7.0		
118	1119-54	11:19:01.7	-54:27:39	290.1	5.9	205.	7.6	0.535783	-14.6	6.5		
119	1133+16	11:33:27.4	+16:07:37	241.9	69.2	5.	0.2	1.187912	-14.4	6.7		
120	1133-55	11:33:38.9	-55:08:32	292.3	5.9	86.	2.9	0.364703	-14.1	5.8		

Notes: B   The pulsar is a member of a binary system.
       M   This is a millisecond pulsar.
       G   This pulsar has displayed glitches in rotation period.

	PSR	RA(1950.0)	DEC(1950.0)	Long (deg)	Lat (deg)	DM	Dist (kpc)	Period (sec)	Pdot	Age	Notes	Association
121	1143-60	11:43:41.7	-60:14:19	295.0	1.3	113.	3.2	0.273372	-14.7	6.4		
122	1154-62	11:54:43.7	-62:08:08	296.7	-0.2	325.	7.0	0.400521	-14.4	6.2		
123	1159-58	11:59:54.1	-58:03:51	296.5	3.9	146.	5.0	0.452801	-14.7	6.5		
124	1221-63	12:21:34.7	-63:51:16	300.0	-1.4	97.	2.7	0.216475	-14.3	5.8		
125	1222-63	12:22:54.5	-63:52:07	300.1	-1.4	416.	14.7	0.419618	-15.0	6.8		
126	1232-55	12:32:31.0	-55:00:00	300.6	7.5	100.	3.5	0.638234				
127	1236-68	12:36:57.0	-68:16:00	301.9	-5.7	96.	3.3	1.301908	-13.9	6.2		
128	1237+25	12:37:12.0	+25:10:17	252.5	86.5	9.	0.3	1.382449	-15.0	7.4		
129	1237-41	12:37:33.9	-41:08:24	300.7	21.4	44.	1.6	0.512242	-14.8	6.7		
130	1240-64	12:40:18.3	-64:06:58	302.1	-1.5	297.	6.0	0.388479	-14.3	6.1		
131	1254-10	12:54:28.1	-10:10:53	305.2	52.4	29.	1.1	0.617308	-15.4	7.4		
132	1256-67	12:56:08.6	-67:25:29	303.7	-4.8	95.	3.2	0.663329	-14.9	6.9		
133	1302-64	13:02:10.9	-64:39:23	304.4	-2.1	506.	19.3	0.571647	-14.4	6.4		
134	1309-53	13:09:03.1	-53:46:47	306.0	8.7	133.	4.8	0.728154	-15.8	7.9		
135	1309-12	13:09:14.5	-12:12:06	310.7	50.1	35.	1.3	0.447518	-15.8	7.7		
136	1309-55	13:09:50.7	-55:00:53	306.0	7.5	134.	4.8	0.849236	-14.2	6.4		
137	1317-53	13:17:49.1	-53:43:24	307.3	8.6	98.	3.5	0.279727	-14.0	5.7		
138	1322-66	13:22:33.2	-66:45:16	306.3	-4.4	211.	7.3	0.543009	-14.3	6.2		
139	1322+83	13:22:39.2	+83:39:18	121.9	33.7	14.	0.5	0.670037	-15.2	7.3		
140	1323-63	13:23:09.2	-63:53:09	306.7	-1.5	505.	18.3	0.792670	-14.5	6.6		
141	1323-58	13:23:44.3	-58:43:56	307.5	3.6	283.	9.8	0.477990	-14.5	6.4		
142	1323-62	13:23:57.1	-62:07:10	307.1	0.2	318.	7.9	0.529906	-13.7	5.6		
143	1325-43	13:25:09.1	-43:42:13	309.9	18.4	42.	1.5	0.532699	-14.5	6.4		
144	1325-49	13:25:31.1	-49:06:03	309.1	13.1	118.	4.3	1.478721	-15.2	7.6		
145	1336-64	13:36:27.5	-64:41:31	308.0	-2.6	77.	2.3	0.378622	-14.3	6.1		
146	1338-62	13:38:30.0	-62:07:00	308.8	-0.1	880.	24.7	0.193202				
147	1352-51	13:52:44.3	-51:39:14	313.0	9.7	112.	4.0	0.644301	-14.6	6.6		
148	1353-62	13:53:50.0	-62:13:00	310.5	-0.6	434.	12.4	0.455761				
149	1356-60	13:56:26.2	-60:23:36	311.2	1.1	295.	8.8	0.127501	-14.2	5.5		
150	1358-63	13:58:10.5	-63:43:17	310.6	-2.1	98.	2.9	0.842785	-13.8	5.9		

Notes: B   The pulsar is a member of a binary system.
       M   This is a millisecond pulsar.
       G   This pulsar has displayed glitches in rotation period.

	PSR	RA(1950.0)	DEC(1950.0)	Long (deg)	Lat (deg)	DM	Dist (kpc)	Period (sec)	Pdot	Age	Notes	Association
151	1359-51	13:59:41.0	-51:10:00	314.1	9.9	39.	1.3	1.380177				
152	1417-54	14:17:02.7	-54:02:39	315.8	6.4	130.	4.5	0.935772	-15.6	7.8		
153	1424-55	14:24:54.7	-55:17:26	316.4	4.8	82.	2.7	0.570290	-14.7	6.6		
154	1426-66	14:26:35.3	-66:09:46	312.7	-5.4	65.	2.1	0.785440	-14.6	6.7		
155	1436-63	14:36:31.5	-63:31:56	314.6	-3.4	124.	4.0	0.459605	-15.0	6.8		
156	1449-64	14:49:25.3	-64:01:00	315.7	-4.4	71.	2.2	0.179484	-14.6	6.0		
157	1451-68	14:51:29.1	-68:31:31	313.9	-8.5	9.	0.2	0.263377	-16.0	7.6		
158	1454-51	14:54:08.6	-51:10:52	322.1	6.7	37.	1.1	1.748301	-14.3	6.7		
159	1503-51	15:03:04.3	-51:46:37	323.1	5.5	61.	1.9	0.840739	-14.2	6.3		
160	1503-66	15:03:23.2	-66:29:26	315.9	-7.3	130.	4.5	0.355655	-14.9	6.7		
161	1504-43	15:04:14.0	-43:40:33	327.3	12.5	49.	1.7	0.286757	-14.8	6.5		
162	1507-44	15:07:27.3	-44:10:47	327.6	11.7	84.	3.0	0.943871	-15.2	7.4		
163	1508+55	15:08:03.8	+55:42:56	91.3	52.3	20.	0.7	0.739678	-14.3	6.4		
164	1509-58	15:09:59.1	-58:56:58	320.3	-1.2	235.	6.7	0.150231	-11.8	3.2		SNR G320.4-1.2
165	1510-48	15:10:44.6	-48:23:11	325.9	7.8	52.	1.7	0.454839	-15.0	6.9		
166	1523-55	15:23:50.3	-55:41:42	323.6	0.6	358.	9.3	1.048705	-13.9	6.2		
167	1524-39	15:24:42.1	-39:21:12	333.1	14.0	49.	1.7	2.417582	-13.7	6.3		
168	1530+27	15:30:04.6	+27:55:57	43.5	54.5	15.	0.5	1.124836	-15.1	7.3		
169	1530-53	15:30:22.5	-53:24:17	325.7	1.9	25.	0.6	1.368881	-14.8	7.2		
170	1540-06	15:40:50.4	-06:11:17	0.6	36.6	19.	0.7	0.709064	-15.1	7.1		
171	1541-52	15:41:12.6	-52:59:23	327.3	1.3	35.	0.9	0.178554	-16.2	7.7		
172	1541+09	15:41:14.4	+09:38:43	17.8	45.8	35.	1.3	0.748448	-15.4	7.4		
173	1550-54	15:50:05.3	-54:47:15	327.2	-0.9	210.	5.6	1.081328	-13.8	6.0		
174	1552-31	15:52:10.0	-31:25:07	342.7	16.8	73.	2.6	0.518110	-16.3	8.2		
175	1552-23	15:52:32.0	-23:33:10	348.4	22.5	51.	1.9	0.532577	-15.2	7.1		
176	1555-55	15:55:23.4	-55:37:09	327.2	-2.0	211.	6.3	0.957242	-13.7	5.9		
177	1556-44	15:56:11.0	-44:30:17	334.5	6.4	59.	1.9	0.257056	-15.0	6.6		
178	1556-57	15:56:14.7	-57:42:47	326.0	-3.7	177.	5.7	0.194454	-14.7	6.2		
179	1557-50	15:57:08.8	-50:35:56	330.7	1.6	270.	7.8	0.192598	-14.3	5.8		
180	1558-50	15:58:33.9	-50:51:45	330.7	1.3	170.	2.5	0.864202	-13.2	5.3		

Notes: B The pulsar is a member of a binary system.
       M This is a millisecond pulsar.
       G This pulsar has displayed glitches in rotation period.

	PSR	RA(1950.0)	DEC(1950.0)	Long (deg)	Lat (deg)	DM	Dist (kpc)	Period (sec)	Pdot	Age	Notes	Association
181	1600-27	16:00:05.0	-27:04:12	347.1	18.8	46.	1.7	0.778312	-14.5	6.6		
182	1600-49	16:00:42.0	-49:01:46	332.2	2.4	141.	4.2	0.327417	-15.0	6.7		
183	1601-52	16:01:25.4	-52:49:26	329.7	-0.5	32.	0.8	0.658013	-15.6	7.6		
184	1604-00	16:04:37.9	-00:24:42	10.7	35.5	11.	0.4	0.421816	-15.5	7.3		
185	1607-13	16:07:55.0	-13:14:34	359.4	26.9	48.	1.8	1.018393	-15.6	7.8		
186	1609-47	16:09:51.1	-47:06:50	334.6	2.8	161.	5.0	0.382376	-15.2	7.0		
187	1612+07	16:12:15.3	+07:45:01	20.6	38.2	21.	0.8	1.206800	-14.6	6.9		
188	1612-29	16:12:45.0	-29:32:38	347.4	15.1	40.	1.4	2.477567	-14.6	7.2		
189	1620-42	16:20:18.1	-42:49:57	338.9	4.6	295.	9.0	0.364590	-15.0	6.8		
190	1620-09	16:20:34.0	-09:01:13	5.3	27.2	70.	2.6	1.276445	-14.5	6.8		
191	1620-26	16:20:34.1	-26:24:58	351.0	16.0	63.	2.2	0.011076	-18.1	8.3	B,M	
192	1630-59	16:30:48.4	-59:48:29	327.7	-8.3	135.	4.6	0.529121	-14.9	6.8		
193	1633+24	16:33:20.2	+24:24:54	43.0	39.9	24.	0.9	0.490506	-15.9	7.8		GC M4
194	1641-45	16:41:10.3	-45:53:39	339.2	-0.2	475.	5.3	0.455055	-13.7	5.6	G	
195	1641-68	16:41:40.7	-68:26:26	321.8	-14.8	43.	1.5	1.785611	-14.8	7.2		
196	1642-03	16:42:24.6	-03:12:31	14.1	26.1	36.	1.3	0.387689	-14.7	6.5		
197	1647-528	16:47:43.3	-52:50:45	334.6	-5.5	164.	5.4	0.890534	-14.7	6.8		
198	1647-52	16:47:46.7	-52:17:56	335.0	-5.2	179.	5.8	0.635056	-14.7	6.7		
199	1648-42	16:48:17.0	-42:40:00	342.5	0.9	525.	13.5	0.844079				
200	1648-17	16:48:38.4	-17:04:19	2.8	16.9	31.	1.0	0.973392	-14.5	6.7		
201	1649-23	16:49:57.0	-23:58:40	357.3	12.5	67.	2.3	1.703740				
202	1657-13	16:57:04.4	-13:00:42	7.5	17.6	59.	2.1	0.640958	-15.2	7.2		
203	1659-60	16:59:47.9	-60:12:43	329.8	-11.4	54.	1.9	0.306323	-15.0	6.7		
204	1700-32	17:00:07.0	-32:51:22	351.6	5.2	105.	3.4	1.211785	-15.2	7.4		
205	1700-18	17:00:55.4	-18:42:06	3.2	13.6	48.	1.7	0.804341	-14.8	6.9		
206	1701-75	17:01:15.7	-75:35:23	316.7	-20.2	37.	1.3	1.191024	-14.7	7.0		
207	1702-19	17:02:41.0	-19:01:34	3.2	13.0	23.	0.7	0.298986	-14.4	6.1		
208	1706-16	17:06:33.2	-16:37:12	5.8	13.7	25.	0.8	0.653050	-14.2	6.2		
209	1707-53	17:07:50.2	-53:46:38	335.7	-8.5	106.	3.6	0.899218	-13.8	6.0		
210	1709-15	17:09:03.6	-15:06:03	7.4	14.0	58.	2.0	0.868804	-15.0	7.1		

Notes: B   The pulsar is a member of a binary system.
       M   This is a millisecond pulsar.
       G   This pulsar has displayed glitches in rotation period.

	PSR	RA(1950.0)	DEC(1950.0)	Long (deg)	Lat (deg)	DM	Dist (kpc)	Period (sec)	Pdot	Age	Notes	Association
211	1717-29	17:17:23.0	-29:30:09	356.5	4.2	43.	1.2	0.620448	-15.1	7.1		
212	1717-16	17:17:31.9	-16:30:32	7.4	11.5	42.	1.4	1.565599	-14.2	6.6		
213	1718-02	17:18:21.0	-02:09:28	20.1	18.9	66.	2.4	0.477715	-16.1	7.9		
214	1718-32	17:18:48.0	-32:05:04	354.6	2.5	126.	3.7	0.477157	-15.2	7.0		
215	1719-37	17:19:35.5	-37:09:06	350.5	-0.5	100.	2.5	0.236169	-14.0	5.5		
216	1726-00	17:26:00.9	-00:05:25	23.0	18.3	38.	1.3	0.386005	-14.9	6.7		
217	1727-47	17:27:55.4	-47:42:21	342.6	-7.7	122.	4.1	0.829724	-12.8	4.9		
218	1729-41	17:29:17.7	-41:26:42	348.0	-4.5	195.	6.1	0.627981	-13.9	5.9		
219	1730-22	17:30:25.0	-22:26:43	4.0	5.7	42.	1.3	0.871683	-14.1			
220	1732-02	17:32:09.4	-02:10:38	21.9	15.9	64.	2.3	0.839386	-15.4	7.5		
221	1732-07	17:32:22.0	-07:22:57	17.3	13.3	74.	2.6	0.419335	-15.1	7.2		
222	1735-32	17:35:38.6	-32:10:16	356.5	-0.5	51.	1.3	0.768499	-13.7	5.7		
223	1736-31	17:36:09.8	-31:29:36	357.1	-0.2	600.	11.7	0.529438	-14.1	5.8		
224	1736-29	17:36:23.6	-29:01:30	359.2	1.1	137.	3.6	0.322881	-12.3	4.3		
225	1737-30	17:37:21.2	-30:14:10	358.3	0.2	150.	3.5	0.606587				
226	1737+13	17:37:49.2	+13:13:29	37.1	21.7	49.	1.8	0.803050	-14.8	6.9		
227	1737-39	17:37:49.5	-39:26:09	350.6	-4.7	159.	5.1	0.512210	-14.7	6.7		
228	1738-08	17:38:38.7	-08:39:06	17.0	11.3	75.	2.6	2.043082	-14.6	7.2		
229	1740-03	17:40:30.0	-03:36:08	21.7	13.4	35.	1.2	0.444644	-14.5	6.3		
230	1740-13	17:40:47.4	-13:50:24	12.7	8.2	115.	3.9	0.405339	-15.3	7.1		
231	1742-30	17:42:42.0	-30:39:02	358.6	-1.0	89.	2.3	0.367421	-14.0	5.7		
232	1745-12	17:45:28.2	-12:59:56	14.0	7.7	100.	3.4	0.394133	-14.9	6.7		
233	1745-56	17:45:31.5	-56:04:25	336.6	-14.3	58.	2.0	1.332310	-14.7	7.0		
234	1747-46	17:47:57.0	-46:56:40	345.0	-10.2	22.	0.7	0.742352	-14.9	7.0		
235	1749-28	17:49:49.2	-28:06:00	1.5	-1.0	51.	1.0	0.562553	-14.1	6.0		
236	1750-24	17:50:25.9	-24:59:00	4.3	0.5	676.	15.0	0.528333	-13.9	5.8		
237	1753+52	17:53:14.7	+52:01:40	79.6	29.6	35.	1.3	2.391396	-14.8	7.4		
238	1753-24	17:53:53.6	-24:35:00	5.0	0.0	363.	7.4	0.670480	-15.5	7.6		
239	1754-24	17:54:37.0	-24:21:40	5.3	0.0	178.	4.0	0.234094	-13.9	5.5		
240	1756-22	17:56:23.0	-22:05:33	7.5	0.8	177.	4.5	0.460969	-14.0	5.8		

Notes: B   The pulsar is a member of a binary system.
       M   This is a millisecond pulsar.
       G   This pulsar has displayed glitches in rotation period.

	PSR	RA(1950.0)	DEC(1950.0)	Long (deg)	Lat (deg)	DM	Dist (kpc)	Period (sec)	Pdot	Age	Notes	Association
241	1757-23	17:57:00.0	-23:43:00	6.1	-0.1	280.	6.1	1.030820				
242	1758-23	17:58:15.0	-23:05:00	6.8	-0.1	1140.	23.4	0.415764				
243	1758-24	17:58:15.0	-24:53:00	5.3	-1.0	250.	6.4	0.124831				
244	1758-03	17:58:44.3	-03:57:55	23.6	9.3	118.	4.0	0.921490	-14.5	6.6		
245	1800-21	18:00:51.1	-21:37:18	8.4	0.1	238.	5.3	0.133588	-12.9	4.2		
246	1802+03	18:02:40.0	+03:06:14	30.4	11.7	79.	2.8	0.218711	-15.0	6.5		
247	1804-12	18:04:00.0	-12:00:00	17.1	4.2	120.	3.8	0.522760				
248	1804-27	18:04:02.0	-27:15:40	3.8	-3.3	313.	8.9	0.827771	-13.9	6.0		
249	1804-08	18:04:53.9	-08:48:10	20.1	5.6	113.	3.7	0.163727	-16.5	8.0		
250	1805-20	18:05:07.0	-20:58:38	9.4	-0.4	607.	13.1	0.918406	-13.8	5.9		
251	1806-21	18:06:14.7	-21:09:48	9.4	-0.7	381.	9.1	0.702413	-14.4	6.5		
252	1806-53	18:06:40.0	-53:38:46	340.3	-15.9	45.	1.6	0.261049	-15.4	7.0		
253	1809-173	18:09:12.6	-17:19:14	13.1	0.5	254.	6.1	1.205362	-13.7	6.0		
254	1809-175	18:09:21.2	-17:34:09	12.9	0.4	535.	11.7	0.538340	-15.0	6.9		
255	1810+02	18:10:22.3	+02:26:08	30.7	9.7	102.	3.5	0.793902	-14.4	6.5		
256	1811+40	18:11:36.0	+40:12:46	67.4	24.0	42.	1.5	0.931088	-14.6	6.8		
257	1813-17	18:13:23.9	-17:30:18	13.4	-0.4	525.	11.7	0.782312	-14.1	6.2		
258	1813-26	18:13:28.0	-26:50:55	5.2	-4.9	129.	4.2	0.592885				
259	1813-36	18:13:43.0	-36:19:11	356.8	-9.4	94.	3.2	0.387017	-14.7	6.5		
260	1814-23	18:14:00.0	-23:13:00	8.5	-3.3	240.	7.2	0.625470				
261	1815-14	18:15:32.9	-14:23:56	16.4	0.6	627.	15.4	0.291488	-14.7	6.4		
262	1817-13	18:17:29.7	-13:47:42	17.2	0.5	780.	20.1	0.921459	-14.3	6.5		
263	1818-04	18:18:13.7	-04:29:05	25.5	4.7	84.	1.5	0.598073	-14.2	6.2		
264	1819-22	18:19:57.0	-22:57:58	9.4	-4.4	123.	3.9	1.874268	-15.2	7.7		
265	1820-14	18:20:03.6	-14:01:40	17.3	-0.2	650.	13.5	0.214771	-15.0	6.6		
266	1820-31	18:20:31.0	-31:08:16	2.1	-8.3	51.	1.7	0.284053	-14.5	6.2		
267	1820-11	18:20:57.0	-11:15:36	19.8	0.9	430.	11.1	0.279829			B	
268	1821-19	18:21:02.8	-19:47:29	12.3	-3.1	224.	6.8	0.189332	-14.3	5.8		
269	1821+05	18:21:04.0	+05:48:47	35.0	8.9	67.	2.3	0.752906	-15.6	7.7		
270	1821-24	18:21:27.4	-24:53:51	7.8	-5.6	120.	3.9	0.003054	-17.8	7.5	M	GC M28

Notes: B   The pulsar is a member of a binary system.
       M   This is a millisecond pulsar.
       G   This pulsar has displayed glitches in rotation period.

PSR	RA(1950.0)	DEC(1950.0)	Long (deg)	Lat (deg)	DM	Dist (kpc)	Period (sec)	Pdot	Age	Notes	Association
271 1821−11	18:21:42.5	−11:20:16	19.8	0.7	603.	15.7	0.435758	−14.4	6.3		
272 1822−14	18:22:11.7	−14:48:31	16.8	−1.0	354.	9.1	0.279182	−13.6	5.3		
273 1822+00	18:22:41.6	+00:02:36	30.0	5.9	54.	1.7	0.778949	−15.1	7.2		
274 1822−09	18:22:46.2	−09:37:31	21.4	1.3	20.	0.5	0.768959	−13.3	5.4		
275 1823−11	18:23:18.1	−11:33:34	19.8	0.3	320.	7.3	2.093135	−14.3	6.8		
276 1823−13	18:23:23.4	−13:36:34	18.0	−0.7	230.	5.8	0.101441	−13.1	4.3		
277 1824−10	18:24:20.0	−10:00:39	21.3	0.8	432.	11.0	0.245757	−15.0	6.6		
278 1826−17	18:26:48.0	−17:52:45	14.6	−3.4	218.	6.7	0.307129	−14.3	5.9		
279 1828−10	18:28:01.1	−11:01:19	20.8	−0.5	173.	4.2	0.405030	−13.2	5.0		
280 1828−60	18:28:44.5	−60:25:19	334.8	−21.2	35.	1.2	1.889436	−15.6	8.0		
281 1829−10	18:29:28.0	−10:26:00	21.5	−0.5	440.	10.5	0.330354	−13.2	5.2		
282 1829−08	18:29:53.4	−08:29:19	23.3	0.3	303.	7.0	0.647279	−14.0	5.2		
283 1830−08	18:30:56.7	−08:29:53	23.4	0.1	411.	8.6	0.085282	−13.4	5.4		
284 1831−03	18:31:04.0	−03:40:55	27.7	2.3	236.	7.1	0.686677	−13.4	5.4		
285 1831−00	18:31:43.3	−00:13:13	30.8	3.7	88.	2.7	0.520954	−16.8	8.8	B	
286 1831−04	18:31:46.5	−04:28:59	27.0	1.7	79.	2.2	0.290106	−15.7	7.4		
287 1832−06	18:32:24.0	−06:45:08	25.1	0.6	475.	11.7	0.305821	−13.4	5.1		
288 1834−04	18:34:08.3	−10:10:44	22.3	−1.4	318.	8.8	0.562706	−13.9	5.9		
289 1834−04	18:34:12.7	−04:39:10	27.2	1.1	233.	6.4	0.354236	−14.8	6.5		
290 1834−06	18:34:32.9	−06:55:41	25.2	0.0	317.	6.8	1.905809	−15.1	7.6		
291 1838−04	18:38:26.8	−04:28:13	27.8	0.3	324.	7.5	0.186145	−14.2	5.7		
292 1839+09	18:39:32.8	+09:09:11	40.1	6.3	49.	1.6	0.381319	−15.0	6.7		
293 1839−04	18:39:48.1	−04:02:58	28.3	0.2	197.	4.6	1.839944	−15.3	7.8		
294 1839+56	18:39:50.3	+56:38:56	86.1	23.8	26.	0.9	1.652861	−14.8	7.2		
295 1841−05	18:41:24.8	−05:41:40	27.1	−0.9	412.	11.1	0.255697	−14.0	5.6		
296 1841−04	18:41:54.4	−04:36:21	28.1	−0.5	125.	3.1	0.991026	−14.4	6.6		
297 1842−02	18:42:08.0	−02:47:50	29.7	0.2	431.	9.8	0.507719	−13.8	5.7		
298 1842+14	18:42:38.5	+14:51:04	45.6	8.1	41.	1.3	0.375463	−14.7	6.5		
299 1842−04	18:42:55.7	−04:37:50	28.2	−0.8	230.	6.0	0.162249	−14.4	5.8		
300 1844−04	18:44:45.0	−04:05:32	28.9	−0.9	143.	3.8	0.597739	−13.3	5.3		

Notes: B The pulsar is a member of a binary system.
       M This is a millisecond pulsar.
       G This pulsar has displayed glitches in rotation period.

PSR	RA(1950.0)	DEC(1950.0)	Long (deg)	Lat (deg)	DM	Dist (kpc)	Period (sec)	Pdot	Age	Notes	Association
301 1845-19	18:45:20.6	-19:55:49	14.8	-8.3	18.	0.5	4.308179	-13.6	6.5		
302 1845-01	18:45:49.0	-01:27:30	31.3	0.0	159.	3.7	0.659428	-14.3	6.3		
303 1846-06	18:46:26.0	-06:40:26	26.8	-2.5	148.	4.5	1.451293	-13.3	5.7		
304 1848+13	18:48:17.5	+13:32:24	45.0	6.3	59.	1.9	0.345581	-14.8	6.6		
305 1848+04	18:48:34.4	+04:14:36	36.7	2.0	112.	3.3	0.284697	-17.8	9.5		
306 1848+12	18:48:54.5	+12:55:59	44.5	5.9	71.	2.3	1.205299	-13.9	6.2		
307 1849+00	18:49:54.0	+00:28:17	33.5	0.0	680.	14.5	2.180173	-13.0	5.6		
308 1851-79	18:51:46.8	-79:55:51	314.3	-27.1	39.	1.4	1.279193	-14.7	7.0		
309 1851-14	18:51:53.9	-14:25:21	20.5	-7.2	130.	4.4	1.146593	-14.4	6.6		
310 1852+10	18:52:06.0	+10:47:00	42.9	4.3	250.	8.4	0.572341				
311 1854+00	18:54:30.0	+00:55:30	34.5	-0.8	90.	2.3	0.356929				
312 1853+01	18:53:36.0	+01:09:00	34.6	-0.5	98.	2.5	0.267399				
313 1855+02	18:55:12.4	+02:08:37	35.6	-0.4	505.	12.8	0.415814	-13.4	5.2		
314 1855+09	18:55:13.7	+09:39:13	42.3	3.1	13.	0.3	0.005362	-19.8	9.7	B,M	White dwarf
315 1857-26	18:57:43.0	-26:04:54	10.3	-13.5	38.	1.3	0.612209	-15.8	7.8		
316 1859+03	18:59:01.9	+03:26:46	37.2	-0.6	403.	10.8	0.655445	-14.1	6.1		
317 1859+01	18:59:02.7	+01:52:19	35.8	-1.4	102.	2.8	0.288219	-14.6	6.3		
318 1859+07	18:59:13.3	+07:12:14	40.6	1.1	251.	7.1	0.643998	-14.6	6.7		
319 1900+05	19:00:15.5	+05:52:01	39.5	0.2	180.	4.3	0.746570	-13.9	6.0		
320 1900+06	19:00:20.0	+06:11:25	39.8	0.3	530.	13.7	0.673501				
321 1900+01	19:00:58.0	+01:31:09	35.7	-2.0	246.	5.0	0.729302	-14.4	6.5		
322 1900-06	19:00:59.0	-06:36:30	28.5	-5.7	196.	6.4	0.431885	-14.5	6.3		
323 1901+10	19:01:40.0	+10:00:00	43.3	1.8	140.	4.1	1.856568				
324 1902-01	19:02:52.7	-01:01:08	33.7	-3.5	225.	7.2	0.643175	-14.5	6.5		
325 1903+07	19:03:28.0	+07:04:42	40.9	0.1	260.	6.0	0.648039	-14.3	6.3		
326 1904+06	19:04:08.9	+06:36:21	40.6	-0.3	473.	11.8	0.267275	-14.7	6.3		
327 1904+12	19:04:49.0	+12:42:00	46.1	2.4	260.	8.4	0.827096				
328 1905+39	19:05:54.7	+39:57:18	70.9	14.2	30.	1.0	1.235757	-15.3	7.6		
329 1906+09	19:06:35.4	+09:11:22	43.2	0.4	250.	6.2	0.830270	-16.0	8.1		
330 1907+00	19:07:01.6	+00:03:04	35.1	-4.0	113.	3.6	1.016945	-14.3	6.5		

Notes: B The pulsar is a member of a binary system.
       M This is a millisecond pulsar.
       G This pulsar has displayed glitches in rotation period.

PSR	RA(1950.0)	DEC(1950.0)	Long (deg)	Lat (deg)	DM	Dist (kpc)	Period (sec)	Pdot	Age	Notes	Association
331 1907+02	19:07:07.7	+02:49:56	37.6	-2.7	172.	5.4	0.989828	-14.3	6.5		
332 1907+10	19:07:27.3	+10:57:08	44.8	1.0	148.	4.0	0.283639	-14.6	6.2		
333 1907+03	19:07:39.7	+03:53:30	38.6	-2.3	79.	2.3	2.330260	-14.3	6.9		
334 1907-03	19:07:52.3	-03:14:51	32.3	-5.7	206.	6.8	0.504603	-14.7	6.6		
335 1907+12	19:07:54.0	+12:26:43	46.2	1.6	274.	8.5	1.441737	-14.1	6.4		
336 1910+10	19:10:30.0	+10:30:00	44.8	0.1	140.	3.4	0.409338				
337 1910+20	19:10:34.1	+20:59:26	54.1	5.0	88.	2.9	2.232964	-14.0	6.5		
338 1911+13	19:11:06.4	+13:55:42	47.9	1.6	144.	4.2	0.521472	-15.1	7.0		
339 1911-04	19:11:15.2	-04:46:00	31.3	-7.1	89.	3.0	0.825934	-14.4	6.5		
340 1911+09	19:11:30.0	+09:30:00	44.0	-0.6	155.	4.0	1.241964				
341 1911+11	19:11:49.1	+11:16:50	45.6	0.2	80.	2.0	0.600997	-15.2	7.2		
342 1913+167	19:13:04.4	+16:41:50	50.6	2.5	66.	1.9	1.616231	-15.4	7.8		
343 1913+10	19:13:07.3	+10:04:32	44.7	-0.6	246.	6.5	0.404538	-13.8	5.6		
344 1913+16	19:13:12.5	+16:01:08	50.0	2.1	167.	5.2	0.059030	-17.1	8.0	B	
345 1914+09	19:14:09.5	+09:46:03	44.6	-1.0	61.	1.6	0.270253	-14.6	6.2		
346 1914+13	19:14:39.7	+13:07:25	47.6	0.5	237.	6.1	0.281840	-14.4	6.1		
347 1915+13	19:15:21.6	+13:48:29	48.3	0.6	95.	2.4	0.194626	-14.1	5.6		
348 1915+22	19:15:36.0	+22:19:00	55.8	4.6	120.	4.1	0.425906				
349 1916+14	19:16:07.1	+14:39:21	49.1	0.9	30.	0.8	1.180884	-12.7	4.9		
350 1917+00	19:17:17.2	+00:16:03	36.5	-6.2	91.	3.0	1.272256	-14.1	6.4		
351 1918+26	19:18:36.3	+26:44:58	60.1	6.0	27.	0.8	0.785522	-16.5	8.5		
352 1918+19	19:18:52.6	+19:43:02	53.9	2.7	154.	5.0	0.821035	-15.0	7.2		
353 1919+14	19:19:06.4	+14:13:32	49.1	0.0	92.	2.2	0.618180	-14.3	6.2		
354 1919+21	19:19:36.2	+21:47:16	55.8	3.5	12.	0.3	1.337301	-14.9	7.2		
355 1919+20	19:19:40.0	+20:00:00	54.2	2.6	70.	2.1	0.760682				
356 1920+20	19:20:08.0	+20:12:16	54.4	2.6	203.	6.7	1.172761				
357 1920+17	19:20:44.0	+21:04:52	55.3	2.9	217.	7.3	1.077919	-14.1	6.3		
358 1921+17	19:21:06.0	+17:00:00	51.7	0.9	135.	3.7	0.547209				
359 1922+20	19:22:30.0	+20:34:06	55.0	2.3	213.	7.0	0.237790	-14.7	6.3		
360 1923+04	19:23:55.6	+04:25:27	41.0	-5.7	102.	3.4	1.074077	-14.6	6.8		

Notes: B   The pulsar is a member of a binary system.
       M   This is a millisecond pulsar.
       G   This pulsar has displayed glitches in rotation period.

	PSR	RA(1950.0)	DEC(1950.0)	Long (deg)	Lat (deg)	DM	Dist (kpc)	Period (sec)	Pdot	Age	Notes	Association
361	1924+19	19:24:16.0	+19:20:00	54.1	1.4	420.	14.3	1.346010		5.7		
362	1924+16	19:24:30.3	+16:42:27	51.9	0.1	177.	4.2	0.579812	-13.7	8.0		
363	1924+14	19:24:39.6	+14:28:48	49.9	-1.0	205.	5.8	1.324922	-15.7			
364	1925+18	19:25:00.0	+18:50:00	53.8	1.0	250.	7.3	0.482765				
365	1925+22	19:25:00.1	+22:28:50	57.0	2.7	180.	6.0	1.431066	-15.1	7.5		
366	1925+188	19:25:27.0	+18:50:00	53.8	0.9	90.	2.4	0.298312				
367	1926+18	19:26:58.0	+18:39:50	53.9	0.5	109.	2.8	1.220469				
368	1927+13	19:27:41.7	+13:09:52	49.1	-2.3	207.	6.6	0.760032	-14.4	6.5		
369	1929+15	19:29:30.0	+15:30:00	51.4	-1.6	120.	3.4	0.314351				
370	1929+10	19:29:51.9	+10:53:04	47.4	-3.9	3.	0.1	0.226517	-14.9	6.5		
371	1929+20	19:29:57.2	+20:14:18	55.6	0.6	211.	5.7	0.268215	-14.4	6.0		
372	1930+22	19:30:14.7	+22:14:30	57.4	1.6	211.	6.6	0.144450	-13.2	4.6		
373	1930+13	19:30:58.0	+13:00:00	49.4	-3.1	165.	5.4	0.928325				
374	1931+24	19:31:00.0	+24:15:00	59.2	2.4	89.	2.7	0.813680				
375	1933+17	19:33:19.0	+17:40:00	53.7	-1.3	210.	6.3	0.654408				
376	1933+16	19:33:31.9	+16:09:58	52.4	-2.1	159.	6.0	0.358736	-14.2	6.0		
377	1933+15	19:33:44.8	+15:29:53	51.9	-2.5	165.	5.3	0.967338	-14.4	6.6		
378	1935+25	19:35:01.5	+25:37:15	60.8	2.3	62.	1.8	0.200980	-14.8	6.3		
379	1937+24	19:37:08.0	+24:43:00	60.3	1.4	100.	2.8	0.645277				
380	1937+21	19:37:28.7	+21:28:01	57.5	-0.3	71.	5.0	0.001558	-19.0	8.4	M	
381	1937-26	19:37:58.0	-26:08:54	13.9	-21.8	50.	1.8	0.402857	-15.0	6.8		
382	1939+17	19:39:47.0	+17:40:00	54.5	-2.7	175.	5.7	0.696261				
383	1940-12	19:40:38.1	-12:44:53	27.2	-17.2	29.	1.0	0.972428	-14.8	7.0		
384	1941-17	19:41:12.0	-17:57:27	22.3	-19.4	56.	2.0	0.841157	-15.0	7.1		
385	1942+17	19:42:15.0	+17:50:00	54.9	-3.1	160.	5.3	1.996898				
386	1942-00	19:42:53.8	-00:48:18	38.6	-12.3	58.	2.0	1.045632	-15.3	7.5		
387	1943+18	19:43:26.0	+18:28:00	55.6	-3.0	215.	7.3	1.068707				
388	1943-29	19:43:44.0	-29:21:02	11.1	-24.1	44.	1.6	0.959447	-14.8	7.0		
389	1944+22	19:44:16.2	+22:37:35	59.3	-1.1	140.	3.9	1.334450	-15.1	7.4		
390	1944+17	19:44:38.8	+17:58:15	55.3	-3.5	16.	0.4	0.440618	-16.6	8.5		

Notes: B   The pulsar is a member of a binary system.
       M   This is a millisecond pulsar.
       G   This pulsar has displayed glitches in rotation period.

PSR	RA(1950.0)	DEC(1950.0)	Long (deg)	Lat (deg)	DM	Dist (kpc)	Period (sec)	Pdot	Age	Notes	Association
391 1946-25	19:46:24.0	-25:32:17	15.2	-23.4	23.	0.8	0.957615	-14.5	6.6		
392 1946+35	19:46:34.0	+35:32:38	70.7	5.0	129.	4.6	0.717307	-14.2	6.2		
393 1949+14	19:49:43.0	+14:03:00	52.5	-6.5	60.	2.0	0.275044				
394 1951+32	19:51:02.5	+32:44:50	68.8	2.8	45.	1.3	0.039530	-14.2	5.0		SNR CTB 80
395 1952+29	19:52:21.8	+29:15:22	65.9	0.8	8.	0.2	0.426677	-17.7	9.5		
396 1953+29	19:53:26.7	+29:00:44	65.8	0.4	105.	2.7	0.006133	-19.5	9.5	B,M	
397 1953+50	19:53:57.4	+50:51:54	84.8	11.6	32.	1.1	0.518937	-14.9	6.8		
398 1957+20	19:57:10.0	+20:40:00	59.2	-4.6	29.	0.8	0.001607			B,M	Dwarf star
399 2000+32	20:00:07.2	+32:08:54	69.3	0.9	143.	4.0	0.696738	-13.0	5.0		
400 2000+40	20:00:59.9	+40:42:27	76.6	5.3	128.	4.6	0.905066	-14.8	6.9		
401 2002+31	20:02:53.7	+31:28:34	69.0	0.0	235.	8.0	2.111217	-13.1	5.7		
402 2003-08	20:03:34.3	-08:15:36	34.1	-20.3	26.	0.9	0.580871	-16.4	8.4		
403 2011+38	20:11:21.6	+38:36:38	75.9	2.5	239.	8.6	0.230191	-14.1	5.6		
404 2016+28	20:16:00.2	+28:30:30	68.1	-4.0	14.	1.3	0.557953	-15.8	7.8		
405 2020+28	20:20:33.3	+28:44:43	68.9	-4.7	25.	1.3	0.343401	-14.7	6.5		
406 2021+51	20:21:25.3	+51:45:07	87.9	8.4	23.	0.7	0.529195	-14.5	6.4		
407 2022+50	20:22:14.1	+50:27:49	86.9	7.5	32.	1.0	0.372618	-14.6	6.4		
408 2025+21	20:25:03.0	+21:35:00	63.5	-9.6	97.	3.5	0.398173				
409 2027+37	20:27:31.2	+37:34:05	76.9	-0.7	189.	5.4	1.216801	-13.9	6.2		
410 2028+22	20:28:28.5	+22:18:13	64.6	-9.8	72.	2.6	0.630512	-15.1	7.1		
411 2034+19	20:34:59.3	+19:32:23	63.2	-12.7	36.	1.2	2.074377	-14.7	7.2		
412 2035+36	20:35:31.8	+36:10:51	76.7	-2.8	92.	2.9	0.618714	-14.3	6.3		
413 2036+53	20:36:38.7	+53:08:37	90.4	7.3	158.	6.2	1.424568	-15.0	7.3		
414 2043-04	20:43:22.4	-04:32:25	42.7	-27.4	36.	1.3	1.546938	-14.8	7.2		
415 2044+15	20:44:19.6	+15:29:30	61.1	-16.8	40.	1.4	1.138286	-15.7	8.0		
416 2045+56	20:45:30.1	+56:57:33	94.2	8.6	99.	3.7	0.476729	-14.0	5.8		
417 2045-16	20:45:47.0	-16:27:53	30.5	-33.1	12.	0.4	1.961567	-14.0	6.5		
418 2048-72	20:48:41.4	-72:12:03	321.9	-35.0	18.	0.6	0.341336	-15.7	7.4		
419 2053+21	20:53:25.0	+21:57:56	67.8	-14.7	36.	1.3	0.815181	-14.9	7.0		
420 2053+36	20:53:33.2	+36:18:50	79.1	-5.6	98.	3.4	0.221508	-15.4	7.0		

Notes: B   The pulsar is a member of a binary system.
       M   This is a millisecond pulsar.
       G   This pulsar has displayed glitches in rotation period.

PSR	RA(1950.0)	DEC(1950.0)	Long (deg)	Lat (deg)	DM	Dist (kpc)	Period (sec)	Pdot	Age	Notes	Association
421 2106+44	21:06:31.7	+44:29:37	86.9	-2.0	140.	4.6	0.414870	-16.1	7.9		
422 2110+27	21:10:54.2	+27:41:38	75.0	-14.0	25.	0.8	1.202851	-14.6	6.9		
423 2111+46	21:11:37.8	+46:31:42	89.0	-1.3	142.	4.3	1.014684	-15.1	7.3		
424 2113+14	21:13:51.1	+14:01:48	64.5	-23.4	56.	2.1	0.440153	-15.5	7.4		
425 2122+13	21:22:23.3	+13:54:22	65.8	-25.1	30.	1.1	0.694053	-15.1	7.2		
426 2123-67	21:23:19.6	-67:01:31	326.4	-39.8	35.	1.3	0.325771	-15.6	7.4		
427 2127+11	21:27:33.1	+11:56:49	65.0	-27.3	58.	2.2	0.110665				GC M15
428 2148+63	21:48:36.8	+63:15:40	104.3	7.4	128.	5.0	0.380140	-15.8	7.6		
429 2148+52	21:48:51.2	+52:33:45	97.5	-0.9	146.	4.3	0.332203	-14.0	5.7		
430 2151-56	21:51:34.0	-56:56:10	337.0	-47.1	14.	0.5	1.373654	-14.4	6.7		
431 2152-31	21:52:18.4	-31:33:09	15.8	-51.6	14.	0.5	1.030002	-14.9	7.1		
432 2154+40	21:54:57.2	+40:03:26	90.5	-11.3	71.	2.6	1.525263	-14.5	6.8		
433 2210+29	22:10:06.8	+29:18:15	86.1	-21.7	73.	2.8	1.004592	-15.3	7.5		
434 2217+47	22:17:45.9	+47:39:48	98.4	-7.6	44.	1.5	0.538467	-14.6	6.5		
435 2224+65	22:24:17.4	+65:20:15	108.6	6.8	35.	1.1	0.682534	-14.0	6.0	G	
436 2227+61	22:27:56.7	+61:50:12	107.2	3.6	123.	4.5	0.443054	-14.6	6.5		
437 2241+69	22:41:23.0	+69:35:07	112.2	9.7	41.	1.4	1.664499	-14.3	6.7		
438 2255+58	22:55:54.2	+58:53:10	108.8	-0.6	151.	4.4	0.368244	-14.2	6.0		
439 2303+30	23:03:34.1	+30:43:49	97.7	-26.7	50.	1.9	1.575885	-14.5	6.9		
440 2303+46	23:03:39.2	+46:51:32	104.9	-12.0	61.	2.3	1.066371	-15.2	7.5	B	
441 2306+55	23:06:02.5	+55:31:20	108.7	-4.2	47.	1.5	0.475068	-15.7	7.6		
442 2310+42	23:10:47.7	+42:36:53	104.4	-16.4	17.	0.6	0.349434	-15.9	7.7		
443 2315+21	23:15:29.0	+21:33:26	95.8	-36.1	21.	0.8	1.444653	-15.0	7.3		
444 2319+60	23:19:41.4	+60:08:02	112.1	-0.6	94.	2.8	2.256484	-14.2	6.7		
445 2321-61	23:21:33.7	-61:10:33	320.4	-53.2	16.	0.6	2.347485	-14.6	7.2		
446 2323+63	23:23:00.2	+63:00:28	113.4	2.0	195.	7.3	1.436308	-14.5	6.9		
447 2324+60	23:24:26.8	+60:55:53	112.9	0.0	123.	3.2	0.233652	-15.5	7.1		
448 2327-20	23:27:49.7	-20:22:04	49.4	-70.2	8.	0.3	1.643620	-14.3	6.7		
449 2334+61	23:34:45.0	+61:34:25	114.3	0.2	58.	1.5	0.495240	-12.7	4.6		
450 2351+61	23:51:34.8	+61:39:07	116.2	0.2	95.	2.5	0.944777	-13.8	6.0		

Notes: B The pulsar is a member of a binary system.
       M This is a millisecond pulsar.
       G This pulsar has displayed glitches in rotation period.

References

Chapter 1

Ables, J. G., Jacka, C. E., Hall, P. J., Hamilton, P. A., McConnell, D. & McCulloch, P. M. (1987) *IAU Circ.* No. 4422.

Baade, W. & Zwicky, F. (1934) *Proc. Natl. Acad. Sci. USA*, **20**, 254.

Boldt, E. A., Desai, U. D., Holt, S. S., Serlemitsos, P. J. & Silverberg, R. F. (1969) *Nature*, **223**, 728.

Bradt, H., Rappaport, S., Mayer, W., Nather, R. E., Warner, B., Macfarlane, M. & Kristian, J. (1969) *Nature*, **222**, 728.

Cocke, W. J., Disney, M. J. & Taylor, D. J. (1969) *Nature*, **221**, 525.

Fishman, G. J., Harnden, F. R. & Haymes, R. C. (1969) *Astrophys. J.*, **156**, L107.

Fritz, G., Henry, R. C., Meekins, J. F., Chubb, T. A. & Friedmann, H. (1969) *Science*, **164**, 709.

Giacconi, R., Gursky, H., Paolini, F. R. & Rossi, B. (1962) *Phys. Rev. Lett.*, **9**, 439.

Giacconi, R., Gursky, H., Kellogg, E., Schreier, E. & Tananbaum, H. (1971) *Astrophys J.*, **167**, L67.

Gold, T. (1968) *Nature*, **218**, 731.

Gold, T. (1969) *Nature*, **221**, 25.

Hayakawa, S. & Matsouka, M. (1964) *Prog. Theor. Phys., Suppl.*, **30**, 204.

Hewish, A., Bell, S. J., Pilkington, J. D. H., Scott, P. F. & Collins, R. A. (1968) *Nature*, **217**, 709.

Hoyle, F., Narlikar, J. & Wheeler, J. A. (1964) *Nature*, **203**, 914.

Large, M. I., Vaughan, A. F. & Mills, B. Y. (1958) *Nature*, **220**, 340.

Melzer, D. W. & Thorne, K. S. (1966) *Astrophys. J.*, **145**, 514.

Middleditch, J. & Pennypacker, C. R. (1985) *Nature*, **313**, 659.

Miller, J. S. & Wampler, E. J. (1969) *Nature*, **221**, 1037.

Oppenheimer, J. & Volkoff, G. M. (1939) *Phys. Rev.*, **55**, 374.

Ostriker, J. P. (1968) *Nature*, **217**, 1127.

Pacini, F. (1967) *Nature*, **216**, 567.

Pacini, F. (1968) *Nature*, **219**, 145.

Pacini, F. & Salpeter, E. E. (1968) *Nature*, **218**, 733.

Radhakrishnan, V. & Cooke, D. J. (1969) *Astrophys. Lett.*, **3**, 225.

References

Radhakrishnan, V. & Manchester, R. M. (1969) *Nature*, **222**, 228.
Reichley, P. E. & Downs, G. S. (1969) *Nature*, **222**, 229.
Richards, D. W. & Comella, J. M. (1969) *Nature*, **222**, 551.
Seward, F. D., Harnden, F. R. & Helfand, D. J. (1984) *Astrophys. J.*, **287**, L19.
Staelin, D. H. & Reifenstein, E. C. (1968) *Science*, **162**, 1481.
Willstrop, R. V. (1969) *Nature*, **221**, 1023.
Zel'dovich, Ya. B. & Guseynov, O. K. (1964) *Astrophys J.*, **144**, 840.

Chapter 2
Arnett, W. D. & Bowers, R. L. (1977) *Astrophys. J. Suppl. Ser.*, **33**, 415.
Baym, G., Pethick, C. & Sutherland, P. (1971) *Astrophys. J.*, **170**, 299.
Fitzpatrick, R. & Mestel, L. (1988) *Mon. Not. R. Astron. Soc.*, **232**, 277 & 303.
Goldreich, P. & Julian, W. H. (1969) *Astrophys. J.*, **157**, 869.
Irvine, J. M. (1978) *Neutron Stars*, Clarendon Press, Oxford.
Mestel, L. (1971) *Nature Phys. Sci.*, **233**, 149.
Ruderman, M. (1974) *IAU Symposium 53*, 117.

Chapter 3
Ables, J. C., Jacka, C. E., Hall, P. J., Hamilton, P. A., McConnell, D. & McCulloch, P. M. (1987) *IAU Circ.* No. 4422.
Backer, D. C. (1987) *IAU Symp. No. 125*, p. 3.
Backer, D. C., Kulkarni, S. R. Heiles, C., Davis, M. M. & Goss, W. M. (1982) *Nature*, **300**, 615.
Clifton, T. R., Backer, D. C., Foster, R. S., Kulkarni, S. R., Fruchter, A. S. & Taylor, J. H. (1987) *IAU Circ.* No. 4422.
Clifton, T. R. & Lyne, A. G. (1986) *Nature*, **320**, 43.
Davies, J. G. & Large, M. I. (1970) *Mon. Not. R. Astron. Soc.*, **149**, 301.
Davies, J. G., Lyne, A. G. & Seiradakis, J. (1972) *Nature*, **240**, 229.
Davies, J. G., Lyne, A. G. & Seiradakis, J. (1973) *Nature Phys. Sci.*, **244**, 84.
Dewey, R. J., Taylor, J. H., Weisberg, J. M. & Stokes, G. H. (1985) *Astrophys. J.*, **294**, L25.
Hulse, R. A. & Taylor, J. H. (1974) *Astrophys. J.*, **191**, L59.
Hulse, R. A. & Taylor, J. H. (1975) *Astrophys. J.*, **201**, L55.
Large, M. I. & Vaughan, A. E. (1971) *Mon. Not. R. Astron. Soc.*, **151**, 277.
Lyne, A. G., Brinklow, A., Middleditch, J., Kulkarni, S. R., Backer, D. C. & Clifton, T. R. (1987) *Nature*, **328**, 399.
Manchester, R. N. (1987) *IAU Symp. No. 125*, p. 13.
Manchester, R. N., Lyne, A. G., Taylor, J. H., Durdin, J. M., Large, M. I. & Little, A. G. (1978) *Mon. Not. R. Astron. Soc.*, **185**, 409.
Manchester, R. N., D'Amico, N. & Tuohy, I. R. (1985) *Mon. Not. R. Astron. Soc.*, **212**, 975.
Slee, O. B., Dulk, G. A. & Otrupek, R. E. (1980) *Proc. Astron. Soc. Aust*, **4**, 100.
Stokes, G. H., Taylor, J. H., Weisberg, J.M. & Dewey, R. J. (1985) *Nature*, **317**, 787.
Strom, R. G. (1987) *Astrophys. J.*, **319**, L103.
Vaughan, A. E. & Large, M. I. (1969) *Proc. Astron. Soc. Aust.*, **1**, 220.

Chapter 4

Ables, J. G. & Manchester, R. N. (1976) *Astron. Astrophys.*, **50**, 177.

Ables, J. G., Jacka, C. E., McConnell, D., Hamilton, P. A., McCulloch, P. M. & Hall, P. J. (1988) *IAU Circ.* No. 4602.

Alexander, J. K., Brown, L. W., Clark, T. A. & Stone, R.G. (1970) *Astron. Astrophys.*, **6**, 476.

Backer, D. C. & Sramek, R. A. (1981) *IAU Symposium 95*, 205.

Backer, D. C. & Sramek, R. A. (1982) *Astrophys. J.*, **260**, 512.

Booth, R. S. & Lyne, A. G. (1976) *Mon. Not. R. Astron. Soc.*, **174**, 53P.

Clifton, T. R., Frail, D. A., Kulkarni, S. R. & Weisberg, J. M. (1988) *Astrophys. J.*, **333**, 332.

Caswell, J. L., Roger, R. S., Murray, J. D., Cole, D. J. & Cooke, D. J. (1975) *Astron. Astrophys.*, **45**, 239.

de Jager, G., Lyne, A. G., Pointon, L. & Ponsonby, J. E.B. (1968) *Nature*, **220**, 128.

Gomez-Gonzales, J. & Guelin, M. (1974) *Astron. Astrophys.*, **32**, 441.

Gordon, K. J. & Gordon, C. P. (1975) *Astron. Astrophys.*, **40**, 27.

Graham, D. A., Mebold, U., Hesse, K. H., Hills, D. Ll. & Wielebinski, R. (1974) *Astron. Astrophys.*, **37**, 405.

Grewing, M. & Warmsley, M. (1971) *Astron. Astrophys.*, **11**, 65.

Gwinn, C. R., Taylor, J. H., Weisberg, J. M. & Rawley, L. A. (1986) *Astron. J.*, **91**, 338.

Kerr, F. J. (1969) *Annu. Rev. Astron. Astrophys.*, **7**, 39.

Lyne, A. G., Brinklow, A., Middleditch, J., Kulkarni, S. R., Backer, D. C. & Clifton, T. R. (1987) *Nature*, **328**, 399.

Lyne, A. G., Biggs, J. D., Brinklow, A., Ashworth, M. & McKenna, J. (1988) *Nature*, **332**, 45.

Lyne, A. G., Manchester, R. N. & Taylor, J. H. (1985) *Mon. Not.R. Astron. Soc.*, **213**, 613.

Manchester, R. N. & Taylor, J. H. (1981) *Astron. J.*, **86**, 1983.

Manchester, R. N., Wellington, K. J. & McCulloch, P. M. (1981) *IAU Symposium 95*, 445.

Manchester, R. N., Tuohy, I. R. & D'Amico, N. (1982) *Astrophys. J.*, **262**, L31.

Milne, D. K. (1970) *Aust. J. Phys.*, **23**, 425.

Prentice, A. J. R. & ter Haar, D. (1969) *Mon. Not. R. Astron. Soc.*, **146**, 425.

Readhead, A. C. S. & Duffett-Smith, P. J., (1975) *Astron. Astrophys*, **42**, 151.

Salter, M. J., Lyne, A. G. & Anderson, B. (1979) *Nature*, **280**, 477.

Strömgren, B. (1936) *Astrophys. J.*, **89**, 526.

Taylor, J. H., Gwinn, C. R., Weisberg, J. M. & Rawley, L. A. (1984) *Very Long Baseline Interferometry, IAU Symposium 110*, 347.

Weisberg, J. M., Boriakoff, V. & Rankin, J. M. (1979) *Astron. Astrophys*, **77**, 204.

Weisberg, J. M., Rankin, J. M. & Boriakoff, V. (1980) *Astron. Astrophys.*, **88**, 84.

Wolszcan, A., Middleditch, J. M., Kulkarni, S. R., Backer, D. C. & Fruchter, A. S. (1988) *IAU Circ.* No. 4552.

Chapter 5

Ash, M. E., Shapiro, I. I. & Smith, W. B. (1967) *Astron. J.*, **72**, 338.
Backer, D. C. & Hellings, R. W. (1986) *Annu. Rev. Astron. Astrophys*, **24**, 527.
Blandford, R. & Teukolsky, S. A. (1976) *Astrophy. J.*, **205**, 580.
Clemence, G. M. & Szebehely, V. (1967) *Astron. J.*, **72**, 1324.
Davis, M. M., Taylor, J. H., Weisberg, J.M. & Backer, D. C. (1985) *Nature*, **315**, 547.
Fomalont, E. B., Goss, W. M., Lyne, A. G. & Manchester, R. N. (1984) *Mon. Not. R. Astr. Soc.*, **210**, 113.
Helfand, D. J., Taylor, J. H. & Manchester, R. N. (1977) *Astrophys. J.*, **213**, L1.
Hulse, R. A. & Taylor, J. H. (1974) *Astrophys. J.*, **195**, L51.
Hunt, G. C. (1971) *Mon. Not. R. Astron. Soc.*, **153**, 119.
Lyne, A. G., Anderson, B. & Salter, M. J. (1982) *Mon. Not. R. Astron. Soc.*, **201**, 503.
Lyne, A. G., Pritchard, R. S. & Smith, F. G. (1988) *Mon. Not. R. Astron. Soc.*, **233**, 667.
Manchester, R. N., Durdin, J. M. & Newton, L.M. (1985) *Nature*, **313**, 374.
Manchester, R. N. & Peterson, B. A. (1989) *Nature*.
Manchester, R. N., Taylor, J. H. & Van, Y. Y. (1974) *Astrophys J.*, **189**, L119.
Standish, E. M. Jr (1982) *Astron. Astrophys.*, **114**, 297.
Taylor, J. H. & Weisberg, J. M. (1982) *Astrophys. J.*, **253**, 908.
Weisberg, J. M. & Taylor, J. H. (1984) *Phys. Rev. Lett.*, **52**, 1348.

Chapter 6

Alpar, M., Anderson, P. W., Pines, D. & Shaham, J. (1981) *Astrophys. J.*, **249**, L29.
Anderson, P. W. & Itoh, N. (1975) *Nature*, **256**, 25.
Backus, P. R., Taylor, J. H. & Damashek, M. (1982) *Astrophys. J.*, **255**, L63.
Baym, G., Pethick, C., Pines, D. & Ruderman, M. (1969) *Nature*, **224**, 872.
Boynton, P. E., Groth, E. J., Hutchinson, D. P., Nanos, G. P. Jr, Partridge, R. B. & Wilkinson, D. T. (1972) *Astrophys. J.*, **175**, 217.
Cordes, J. M., Downs, G. S. & Krauss-Polstorff, J. (1988) *Astrophys. J.*, **330**, 847.
Cordes, J. M. & Downs, G. S. (1985) *Astrophys. J., Suppl. Ser.*, **59**, 343.
Cordes, J. M. & Helfand, D. J. (1980) *Astrophys. J.*, **239**, 640.
Demianski, M. & Proszynski, M. (1983) *Mon. Not. R. Astron. Soc.*, **202**, 437.
Downs, G. S. (1982) *Astrophys. J.*, **257**, L67.
Flanagan, C. (1988) *IAU Circ.* No. 4695.
Gullahorn, G. E., Payne, R. R., Rankin, J. M. & Richards, D. W. (1976) *Astrophys. J.*, **205**, L151.
Helfand, D. J., Taylor, J. H., Backus, P. R. & Cordes, J. M. (1980) *Astrophys. J.*, **237**, 206.
Lohsen, E. (1975) *Nature*, **258**, 689.
Lyne, A. G. (1987) *Nature*, **326**, 569.
Lyne, A. G. & Pritchard, R. S. (1987) *Mon. Not. R. Astron. Soc.*, **229**, 223.

Lyne, A. G., Pritchard, R. S. & Smith, F. G. (1988) *Mon. Not. R. Astron. Soc.*, **233**, 667.
McCulloch, P. M., Hamilton, P. A., Royle, G. W. R. & Manchester, R. N. (1983) *Nature*, **302**, 319.
McCulloch, P. M. Klekociuk, A. R., Hamilton, P. A., & Royle, G. W. R. (1987) *Aust. J. Phys.*, **40**, 727.
Manchester, R. N., Newton, L. M., Goss, W. M. & Hamilton, P. A. (1978) *Mon. Not. R. Astron. Soc.*, **184**, 35P.
Manchester, R. N., Newton, L. M., Hamilton, P. A. & Goss, W. M. (1983) *Mon. Not. R. Astron. Soc.*, **202**, 269.
Manchester, R. N. & Taylor, J. H. (1974) *Astrophys. J.*, **191**, L63.
Newton, L. M., Manchester, R. N. & Cooke, D. J. (1981) *Mon. Not. R. Astron. Soc.*, **194**, 841.
Radhakrishnan, V. & Manchester, R. N. (1969) *Nature*, **222**, 228.
Reichley, P. E. & Downs, G. S. (1969) *Nature*, **222**, 229.
Reichley, P. E. & Downs, G. S. (1971) *Nature Phys. Sci.*, **234**, 48.
Ruderman, M. (1970) *Nature*, **225**, 619.
Tkachenko, V. K. (1966) *Soviet Phys. JETP*, **23**, 1049.

Chapter 7
Argyle, E. & Gower, J. F. R. (1972) *Astrophys J.*, **175**, L89.
Bignami, G. F. & Caraveo, P. A. (1988) *Astrophys. J.*, **325**, L5.
Bovkun, V. P. (1979) *Astron. Zh. USSR*, **56**, 699; *Sov. Astron.*, **23**, 394.
Braun, R., Goss, W. M. & Lyne, A. G. (1989) *Astrophys. J.*, **340**, 355.
Buccheri, R. (1981) *IAU Symposium 95*, 241.
Caswell, J. L., Milne, D. K. & Wellington, K. J. (1981) *Mon. Not. R. Astron. Soc.*, **195**, 89.
Clifton, T. R. & Lyne, A. G. (1986) *Nature*, **320**, 43.
Demianski, M. & Prozynski, M. (1983) *Mon. Not. R. Astron. Soc.*, **202**, 437.
Downs, G. S. (1981) *Astrophys. J.*, **249**, 687.
Drake, F. (1971) *IAU Symposium 46*, 73.
Gibson, A. I., Harrison, A. B., Kirkman, I. W., Lotts, A. P., Macrae, H. J., Orford, K. J., Turver, K. E. & Walmsley, M. (1982) *Nature*, **296**, 866.
Hankins, T. H. & Rickett, B. J. (1975) *Methods in Comput. Phys.*, **14**, 55.
Hegyi, D., Novick, R. & Thaddeus, P. (1971) *IAU Symposium 46*, 87.
Heiles, C. & Rankin, J. M. (1971) *Nature Phys. Sci.*, **231**, 97.
Hewish, A. & Okoye, S. E. (1964) *Nature*, **203**, 171.
Jones, D. H. P., Smith, F. G. & Wallace, P. T. (1981) *Mon. Not. R. Astron. Soc.*, **196**, 943.
Kanbach, G., Bennett, K., Bignami, G. F., Buccheri, R., Caraveo, P., D'Amico, N., Hermesen, W., Lichti, G. G., Masnou, J. L., Mayer-Hasselwander, H. A., Paul, J. A., Sacco, B., Swanenburg, B. N. & Wills, R. D. (1980) *Astron. Astrophys.*, **90**, 163.
Large, M. I., Vaughan, A. E. & Mills, B. Y. (1968) *Nature*, **220**, 340.
Lyne, A. G., Pritchard, R. S. & Smith, F. G. (1988) *Mon. Not. R. Astron. Soc.*, **233**, 667.
Lyne, A. G. & Thorne, D. J. (1975) *Mon. Not. R. Astron. Soc.*, **172**, 197.

References

McCulloch, P. M., Hamilton, P. A., Royle, G. W. R. & Manchester, R. N. (1983) *Nature*, **302**, 319.
Manchester, R. N. & Durdin, J. M. (1984) *Bull. Am. Astron. Soc.*, **16**, 542.
Manchester, R. N., Wallace, P. T., Peterson, B. A. & Elliott, K. H. (1980) *Mon. Not. R. Astron. Soc.*, **190**, 9P.
Middleditch, J. & Pennypacker, C. (1985) *Nature*, **313**, 659.
Middleditch, J., Pennypacker, C. & Burne, M. S. (1983) *Astrophys. J.*, **273**, 261.
Pacini, F. (1967) *Nature*, **216**, 567.
Radhakrishnan, V., Cooke, D. J., Komesaroff, M. M. & Morris, D. (1969) *Nature*, **221**, 443.
Rankin, J. M., Campbell, D. B., Isaacman, R. B. & Payne, R. R. (1988) *Astron. Astrophys*, **202**, 166.
Rankin, J. M. & Counselman, C. C. (1973) *Astrophys. J.*, **181**, 875.
Rankin, J. M., Payne, R. R. & Campbell, D. B. (1974) *Astrophys. J.*, **193**, L71.
Rickett, B. J., Coles, Wm. A. & Bourgois, G. (1983) *Astron. Astrophys*, **134**, 390.
Seward, F. D., Harnden, F. R., Murdin, P. & Clark, D. H. (1983) *Astrophys J.*, **267**, 698.
Seward, F. D., Harnden, F. R. & Helfand, D. J. (1984) *Astrophys. J.*, **287**, L19.
Smith, F. G., Jones, D. H. P., Dick, J. B. & Pike, C. D. (1988) *Mon. Not. R. Astron. Soc.*, **233**, 305.
Staelin, D. M. & Reifenstein, E. C. (1968) *Science*, **162**, 1481.
Wallace, P. T., Peterson, B. A., Murdin, P. G., Danziger, I. J., Manchester, R. N., Lyne, A. G., Goss, W. M., Smith, F. G., Disney, M. J., Hartley, K. F., Jones, D. H. P. & Wellgate, G. W. (1977) *Nature*, **266**, 692.
Wilson, R. B. & Fishman, G. J. (1983) *Astrophys. J.*, **269**, 273.

Chapter 8
Blaauw, A. (1985) *Astrophys. & Spa. Sci. Library* (publ. D. Reidel), **120**, 211.
Clifton, T. R. & Lyne, A. G. (1986) *Nature*, **320**, 43.
Damashek, M., Taylor, J. H. & Hulse, R. A. (1978) *Astrophys. J.*, **225**, L31.
Davies, J. G., Lyne, A. G. & Seiradakis, J. (1972) *Nature*, **240**, 229.
Davies, J. G., Lyne, A. G. & Seiradakis, J. (1973) *Nature Phys. Sci.*, **244**, 84.
Dewey, R. J., Taylor, J. H., Weisberg, J. M. & Stokes, G. H. (1985) *Astrophys. J.*, **294**, L25.
Gunn, J. E. & Ostriker, J. P. (1970) *Astrophys. J.*, **160**, 979.
Hulse, R. A. & Taylor, J. H. (1974) *Astrophys. J.*, **191**, L59.
Hulse, R. A. & Taylor, J. H. (1975) *Astrophys. J.*, **201**, L55.
Large, M. I. & Vaughan, A. E. (1971) *Mon. Not. R. Astron. Soc.*, **151**, 277.
Lyne, A. G., Anderson, B. & Salter (1981) *Mon. Not. R. Astron. Soc.*, **201**, 503.
Lyne, A. G., Manchester, R. N & Taylor, J. H. (1985) *Mon. Not. R. Astron. Soc.*, **213**, 613.
Manchester, R. N., Lyne, A. G., Taylor, J. H., Durdin, J. M., Large, M. I. & Little, A. G. (1978) *Mon. Not. R. Astron. Soc.*, **185**, 409.
Manchester, R. N., Taylor, J. H. & Van, Y. Y. (1974) *Astrophys. J.*, **189**, L119.
Stokes, G. H., Taylor, J. H., Weisberg, J. M. & Dewey, R. J. (1985) *Nature*, **317**, 787.

Chapter 9

Baade, W. (1942) *Astrophys. J.*, **96**, 188.
Baade, W. & Zwicky, F. (1934) *Proc. Nat. Acad. Sci.*, **20**, 254.
Baan, W. A. (1982) *Astrophys. J.*, **261**, L71.
Batten, A. H. (1967) *Annu. Rev. Astron. Astrophys.*, **5**, 25.
Blaauw, A. (1961) *Bull. Astron. Inst. Netherlands*, **15**, 265.
Boersma, J. (1961) *Bull. Astron. Inst. Netherlands*, **15**, 291.
Cline, T. L., Desai, U. D. (& *14 others*) (1982) *Astrophys. J.*, **255**, L45.
Colgate, S. A. (1970) *Nature*, **225**, 247.
de Vacouleurs, G. & Corwin, H. G. (1985) *Astrophys. J.*, **295**, 287.
Dombrovsky, V. A. (1954) *Dokl. Akad. Nauk. USSR*, **94**, 1021.
Duncan, J. C. (1939) *Astrophys. J.*, **89**, 482.
Duyvendak, J. J. L. (1942) *Publ. Astron. Soc. Pacific*, **54**, 91.
Ho Peng-Yoke (1962) *Vistas in Astronomy*, ed. A. Beer, p. 127. (Pergamon Press).
Ilovaisky, S. A. & Lequeux, J. (1972) *Astron. Astrophys.*, **18**, 169.
Katgert, P. & Oort, J. H. (1967) *Bull. Astron. Inst. Netherlands*, **19**, 239.
Klebesadel, R. W., Strong, I. B. & Olson, R. A. (1973) *Astrophys. J.*, **182**, L85.
Lampland, C. O. (1921) *Publ. Astron. Soc. Pacific*, **33**, 79.
Lundmark, K. E. (1920) *K. Svenska Vetenskaps. Handl.*, **60**, 53.
Lundmark, K. (1921) *Publ. Astron. Soc. Pacific*, **33**, 225.
Maza, J. & van den Bergh, S. (1976) *Astrophys. J.*, **204**, 519.
McCluskey, G. E. & Kondo, Y. (1971) *Astrophys. Spa. Sci.*, **10**, 464.
Minkowski, R. (1964) *Annu. Rev. Astron. Astrophys.*, **2**, 247.
Oort, J. H. & Walraven, Th. (1956) *Bull. Astron. Inst. Netherlands*, **12**, 285.
Scargle, J. & Harlan, E. (1970) *Astrophys. J.*, **159**, L143.
Shklovsky, I. S. (1953) *Dokl. Akad. Nauk. USSR*, **90**, 983.
van den Bergh, S., McClure, R. D. & Evans, R. (1987) *Astrophys. J.*, **323**, 44.
van den Heuvel, E. P. J. & De Loore, C. (1973) *Astron. Astrophys.*, **25**, 387.
Vashakidze, M. A. (1954) *Astr. Circ.*, No. 147.
Whelan, J. & Iben, I. (1973) *Astrophys. J.*, **186**, 1007.
Woltjer, L. (1972) *Annu. Rev. Astron. Astrophys.*, **10**, 129.
Woltjer, L. (1987) *NATO Adv. Sci. Institutes* (publ. D. Reidel), **195**, 209.

Chapter 10

Backer, D. C., Kulkarni, S. R., Heiles, C., Davis, M. M. & Goss, W. M. (1982) *Nature*, **300**, 615.
Boriakoff, V., Buccheri, R. & Fanti, F. (1983) *Nature*, **304**, 417.
Damashek, M., Backus, P. R., Taylor, J. H. & Burkhardt, R. K. (1982) *Astrophys. J.*, **253**, L57.
Dewey, R. J., Maguire, C. M., Rawley, L. A., Stokes, G. H. & Taylor, J. H. (1986) *Nature*, **322**, 712.
Fruchter, A. S., Stinebring, D. R. & Taylor, J. H. (1988) *Nature*, **333**, 237.
Kulkarni, S. R. (1986) *Astrophys. J.*, **306**, L95.
Lyne, A. G. (1984) *Nature*, **310**, 300.
Lyne, A. G., Brinklow, A., Middleditch, J., Kulkarni, S. R., Backer, D. C. & Clifton, T. R. (1987) *Nature*, **328**, 399.

References

Lyne, A. G., Biggs, J. D., Brinklow, A., Ashworth, M. & McKenna, J. (1988) *Nature*, **332**, 45.
McKenna, J. & Lyne, A. G. (1988) *Nature*, **336**, 226.
Manchester, R. N., Newton, L. M., Cooke, D. J., Backus, P. R., Damashek, M., Taylor, J. H. & Condon, P. J. (1983) *Astrophys. J.*, **268**, 832.
Manchester, R. N., Newton, L. M., Cooke, D. J. & Lyne, A. G. (1980) *Astrophys. J.*, **236**, L25.
Segelstein, D. J., Rawley, L. A., Stinebring, D. A., Fruchter, A. S. & Taylor, J. H. (1986) *Nature*, **322**, 714.
Stokes, G. H., Taylor, J. H. & Dewey, R. J. (1985) *Astrophys. J.*, **294**, L21.
van den Heuvel, E. P. J. (1987) IAU Symposium No. 125, p. 383.
van den Heuvel, E. P. J. & Bonsema, P. T. J. (1984) *Astron. Astrophys.*, **139**, L16.
van den Heuvel, E. P. J. & Taam, R. E. (1984) *Nature*, **309**, 235.
Wolszcan, A., Middleditch, J. M., Kulkarni, S. R., Backer, D. C. & Fruchter, A. S. (1988) *IAU Circ.* No. 4552.
Wright, G. A. & Loh, E. D. (1986) *Nature*, **324**, 127.

Chapter 11
Baan, W. A. (1982) *Astrophys. J.*, **261**, L71.
Belian, R. D., Conner, J. P. & Evans, W. D. (1976) *Astrophys. J.*, **206**, L135.
Bradt, H. V. D. & McClintock, J. E. (1983) *Annu. Rev. Astron. Astrophys.*, **21**, 13.
Giacconi, R. (1974) *Proc. 16th Int. Solvay Conf. on Physics*, (Editions de l'Univ. de Bruxelles), p. 27.
Grindley, J., Gursky, H., Schnopper, H., Parsignault, D. R., Heise, J., Brinkman, A. C. & Shrijver, J. (1976) *Astrophys. J.*, **205**, L127.
Joss, P. C. & Rappaport, S. A. (1984) *Annu. Rev. Astron. Astrophys.*, **22**, 537.
Hernquist, L. (1985) *Mon. Not. R. Astron. Soc.*, **213**, 313.
Lewin, W. H. G. & Joss, P. C. (1981) *Space Sci. Rev.*, **28**, 3.
Lewin, W. H. G., Doty, J., Clark, G. W., Rappaport, S. A., *et al.* (1976) *Astrophys. J.*, **207**, L95.
Lewin, W. H. G. & van Paradijs, J. (1986) *Comments Astrophys.*, **11**, 127.
Primini, F., Rappaport, S. & Joss, P. C. (1977) *Astrophys. J.*, **217**, 543.
Schreier, E., Levinson, R., Gursky, H., Kellogg, E., Tananbaum, H. & Giacconi, R. (1972) *Astrophys. J.*, **172**, L79.
Shklovsky, I. S. (1967) *Astrophys. J.*, **148**, L1.
Trümper, J., Pietsch, W., Reppin, C., Voges, W., Staubert, R. & Kendziorra, E. (1978) *Astrophys. J.*, **219**, L105.
White, N. E., Swank, J. H. & Holt, S. S. S. (1983) *Astrophys. J.*, **270**, 711.

Chapter 12
Backer, D. C. & Rankin, J. A. (1986) *Astrophys. J., Suppl. Ser.*, **42**, 143.
Kuzmin, A. D., Malofeev, V. M., Izvekova, V. A., Sieber, W. & Wielebinski, R. (1986) *Astron. Astrophys.*, **161**, 183.
Lyne, A. G., Smith, F. G. & Graham, D. A. (1971) *Mon. Not. R. Astron. Soc.*, **153**, 337.

Lyne, A. G. & Manchester, R. N. (1988) *Mon. Not. R. Astron. Soc.*, **234**, 477.
Radhakrishnan, V. & Cooke, D. J. (1969) *Astrophys. Lett.*, **3**, 225.
Rankin, J. M. (1983) *Astrophys. J.*, **274**, 333.
Sieber, W. (1973) *Astron. Astrophys.*, **28**, 237.

Chapter 13
Backer, D. C. (1970) *Nature*, **228**, 42.
Bartel, N., Morris, D., Sieber, W. & Hankins, T. H. (1982) *Astrophys. J.*, **258**, 776.
Bartel, N., Sieber, W. & Wolszcan, A. (1980) *Astron. Astrophys*, **90**, 58.
Biggs, J. D., Hamilton, P. A., McCulloch, P. M. & Lyne, A. G. (1985) *Mon. Not. R. Astron. Soc.*, **214**, 47P.
Biggs, J. D., McCulloch, P. M., Hamilton, P. A., Manchester, R. N. & Lyne, A. G. (1985) *Mon. Not. R. Astron. Soc.*, **215**, 281.
Boriakoff, V., Ferguson, D. C. & Slater, G. (1981) *IAU Symp. No. 95*, 199.
Cole, T. W. (1970) *Nature*, **227**, 788.
Cordes, J. M. (1975) *Astrophys. J.*, **195**, 193.
Durdin, J. M., Large, M. I., Manchester, R. N., Lyne, A. G. & Taylor, J. H. (1979) *Mon. Not. R. Astron. Soc.*, **186**, 39P.
Fillipenko, A. V. & Radhakrishnan, V. (1982) *Astrophys. J.*, **263**, 828.
Hankins, T. H. & Cordes, J. M. (1981) *Astrophys. J.*, **249**, 241.
Huguenin, G. R., Taylor, J. H. & Troland, T. H. (1970) *Astrophys. J.*, **162**, 727.
Lyne, A. G. & Ashworth, M. (1983) *Mon. Not. R. Astron. Soc.*, **204**, 519.
Manchester, R. N., Tademaru, E., Taylor, J. H. & Huguenin, G. R. (1973) *Astrophys. J.*, **185**, 951.
Manchester, R. N., Taylor, J. H. & Huguenin, G. R. (1975) *Astrophys. J.*, **196**, 83.
Rickett, B. J., Hankins, T. H. & Cordes, J. M. (1975) *Astrophys. J.*, **201**, 425.
Ritchings, R. T. (1976) *Mon. Not. R. Astron. Soc.*, **176**, 249.
Ritchings, R. T. & Lyne, A. G. (1975) *Nature*, **257**, 293.
Ruderman, M. A. & Sutherland, P. G. (1975) *Astrophys. J.*, **196**, 51.
Taylor, J. H. & Huguenin, G. R. (1971) *Astrophys. J.*, **167**, 273.
Taylor, J. H., Huguenin, G. R. & Hirsch, R. M. (1971) *Astrophys. Lett.*, **9**, 205.
Schonhardt, P. & Sieber, W. (1973) *Astrophys. Lett.*, **14**, 61.
Sieber, W. & Oster, L. (1975) *Astron. Astrophys.*, **38**, 325.
Smith, F. G. (1973) *Mon. Not. R. Astron. Soc.*, **161**, 9P.
Unwin, S. C., Readhead, A. C. S., Wilkinson, P. N. & Ewing, M. S. (1978) *Mon. Not. R. Astron. Soc.*, **182**, 711.
Wright, G. A. E. & Fowler, L. A. (1981) *IAU Symposium No. 95*, 211.

Chapter 14
Cheng, K. S., Ho, C. & Ruderman, M. A. (1986) *Astrophys. J.*, **300**, 500.
Komesaroff, M. M. (1970) *Nature*, **225**, 612.
Smith, F. G. (1986) *Mon. Not. R. Astron. Soc.*, **219**, 729.
Smith, F. G., Jones, D. H. P., Dick, J. B. & Pike, C. D. (1988) *Mon. Not. R. Astron. Soc.*, **233**, 305.

References 269

Chapter 15

Epstein, R. I. (1973) *Astrophys. J.*, **183**, 593.
Ginzburg, V. L. & Syrovatskii, S. I. (1965) *Annu. Rev. Astron. Astrophys.*, **3**, 297.
Ginzburg, V. L. & Syrovatskii, S. I. (1969) *Annu. Rev. Astron. Astrophys.*, **7**, 375.
Ginzberg, V. L. & Zhelesnyakov, V. V. (1970) *Comments Astrophys. Space Phys.*, **2**, 197.
Goldreich, P. & Keeley, D. A. (1971) *Astrophys. J.*, **170**, 463.
Jackson, J. D. (1962) *Classical Electrodynamics*. John Wiley, New York, p. 467.
Komesaroff, M. M. (1970) *Nature*, **225**, 612.
Smith, F. G. (1970) *Mon. Not. R. Astron. Soc.*, **149**, 1.
Smith, F. G. (1971) *Mon. Not. R. Astron. Soc.*, **154**, 5P.
Zhelesnyakov, V. V. (1971) *Astrophys. Space Sci.*, **13**, 87.

Chapter 16

Beskin, V. S., Gurevich, A. V. & Istomin, Ya. N. (1988) *Astrophys. Space Sci.*, **146**, 205.
Cheng, K. S., Ho, C. & Ruderman, M. (1981) *Astrophys. J.*, **300**, 522.
Davila, J., Wright, C. & Benford, G. (1980) *Astrophys. Space Sci.*, **71**, 51.
Filippenko, A. V. & Radhakrishnan, V. (1982) *Astrophys. J.*, **263**, 828.
Knight, F. K. (1982) *Astrophys. J.*, **260**, 538.
Middleditch, J., Pennypacker, C. & Burne, M. S. (1983) *Astrophys. J.*, **273**, 261.
Oke, J. B. (1969) *Astrophys. J.*, **156**, L49.
Shklovsky, I. S. (1970) *Astrophys. J.*, **159**, L77.
Sturrock, P. A. (1971) *Astrophys. J.*, **164**, 529.
Zhelesnyakov, V. V. & Shaposhnikov, V. E. (1972) *Astrophys. Space Sci.*, **18**, 166.

Chapter 17

Ables, J. G., Komesaroff, M. M. & Hamilton, P. A. (1973) *Astrophys. Lett.*, **6**, 147.
Cohen, M. H., Gundermann, E. J., Handebeck, H. E. & Sharp, L. E. (1967) *Astrophys. J.*, **147**, 449.
Cordes, J. M., Weisberg, J. M. & Boriakoff, V. (1985) *Astrophys. J.*, **288**, 221.
Duffett-Smith, P. J. & Readhead, A. C. S. (1976) *Mon. Not. R. Astron. Soc.*, **174**, 7.
Galt, J. A. & Lyne, A. G. (1972) *Mon. Not. R. Astron. Soc.*, **158**, 281.
Hewish, A. (1980) *Mon. Not. R. Astron. Soc.*, **192**, 799.
Lang, K. R. (1971) *Astrophys. J.*, **164**, 249.
Lee, L. C. (1976) *Astrophys. J.*, **206**, 744.
Lee, L. C. & Jokipij, J. R. (1976) *Astrophys. J.*, **206**, 735.
Lyne, A. G. (1971) *IAU Symposium No. 46*, D. Reidel, Dordrecht, p. 182.
Lyne, A. G. & Rickett, B. R. (1968) *Nature*, **218**, 326.
Lyne, A. G. & Smith, F. G. (1982) *Nature*, **298**, 825.

Lyne, A. G. & Thorne, D. J. (1975) *Mon. Not. R. Astron. Soc.*, **172**, 97.
Mutel, R. L., Broderick, J. J., Carr, T. D., Lynch, M., Desch, M., Warnock, W. W. & Klemperer, W. K. (1974) *Astrophys. J.*, **193**, 279.
Rickett, B. J. (1969) *Nature*, **221**, 158.
Rickett, B. J. (1977) *Annu. Rev. Astron. Astrophys.*, **15**, 479.
Rickett, B. J., Coles, Wm. A. & Bourgois, G. (1984) *Astron. Astrophys.*, **134**, 390.
Roberts, J. A. & Ables, J. G. (1982) *Mon. Not. R. Astron. Soc.*, **201**, 1119.
Scheuer, P. A. G. (1968) *Nature*, **218**, 920.
Sieber, W. (1982) *Astron. Astrophys.*, **113**, 311.
Smith, F. G. & Wright, N. C. (1985) *Mon. Not. R. Astron. Soc.*, **214**, 97.
Sutton, J. M., Staelin, D. H. & Price, R. M. (1971) *IAU Symposium No. 46*, D. Reidel, Dordrecht, p. 46.
Uscinski, B. J. (1968) *Philos. Trans. R. Soc. London, Ser. A*, **262**, 609.
Uscinski, B. J. (1974) *Proc. R. Soc. London, Ser. A*, **336**, 379.
Vandenberg, N. R., Clark, T. A., Erickson, W. C., Resch, G. M., Broderick, J. J., Payne, R. R., Knowles, S. H. & Youmans, A. B. (1973) *Astrophys. J.*, **180**, L27.
Wilkinson, P. N., Spencer, R. E. & Nelson, R. F. (1988) *IAU Symposium No. 129*, D. Reidel, Dordrecht, p. 305.
Williamson, I. P. (1973) *Mon. Not. R. Astron. Soc.*, **163**, 345.
Williamson, I. P. (1974) *Mon. Not. R. Astron. Soc.*, **166**, 499.
Wolszcan, A. (1982) *Mon. Not. R. Astron. Soc.*, **204**, 591.

Chapter 18
Beuermann, K., Kanbach, G. & Berkhuijsen, E. M. (1985) *Astron. Astrophys.*, **153**, 17.
Bingham, R. G. & Shakeshaft, J. R. (1967) *Mon. Not. R. Astron. Soc.*, **136**, 347.
Gardner, F. F., Morris, D. & Whiteoak, J. B. (1969) *Aust. J. Phys.*, **22**, 813.
Hall, J. S. & Mikesell, A. H. (1950) *Publ. U.S. Nav. Obs.*, **17**, 1.
Hamilton, P. N. & Lyne, A. G. (1987) *Mon. Not. R. Astron. Soc.*, **224**, 1073.
Hiltner, W. A. (1949) *Astrophys. J.*, **114**, 241.
Lyne, A. G. & Smith, F. G. (1968) *Nature*, **218**, 124.
Lyne, A. G., Smith, F. G. & Graham, D. A. (1971) *Mon. Not. R. Astron. Soc.*, **153**, 337.
Manchester, R. N. (1974) *Astrophys. J.*, **188**, 637.
Simard-Normandin, M. & Kronberg, P. P. (1980) *Astrophys. J.*, **242**, 74.
Smith, F. G. (1968) *Nature*, **218**, 325.
Spoelstra, T. A. T. (1984) *Astron. Astrophys.*, **135**, 238.
van der Hulst, H. C. (1967) *Annu. Rev. Astron. Astrophys.*, **5**, 167.
Wilkinson A. & Smith, F. G. (1974) *Mon. Not. R. Astron. Soc.*, **167**, 593.

Chapter 19
Fiedler, R. L., Dennison, B. K., Johnston, K. J. & Hewish, A. (1987) *Nature*, **326**, 675.
Jones, P. B. (1987) *Mon. Not. R. Astron. Soc.*, **228**, 513.

Lyne, A. G. & Smith, F. G. (1989) *Mon. Not. R. Astron. Soc.*, **237**, 533.
Middleditch, J., Pennypacker, C., Morris, D. E., Muller, R. A., Perlmutter, S., Sasseen, T., Kristian, J. A., Kunkel, W. E., Hamay, M. A., Inamura, J. N., Steiman-Cameron, T. Y., Shelton, I. K. Tuohy, I. R. and Rawlings, S. (1989) *IAU Circ.*, 4735.
van den Heuvel, E. P. J. & Bonsema, P. T. J. (1984) *Astron. Astrophys*, **139**, L16.
van den Heuvel, E. P. J. & van Paradjis, J. (1988) *Nature*, **334**, 227.
Vivekenand, M. & Narayan, R. (1981) *J. Astrophys. Astron.*, **2**, 315.

Chapter 20
Manchester, R. N. & Taylor, J. H. (1981) *Astron. J.*, **86**, 1953.
Taylor, J. H. & Manchester, R. N. (1975) *Astron. J.*, **80**, 794.
Turtle, A. J. & Vaughan, A. E. (1968) *Nature*, **219**, 689.

Index

accretion, 139, 146, 150
accretion-induced collapse, 118, 139
age of pulsars, 52, 80, 107, 128
    apparent, due to proper motion, 57
    characteristic, 53, 105, 109, 128
    compared with supernova remnants, 110, 121, 240
    Crab Pulsar, 67, 122
Alfvén surface, 147
alignment of magnetic and rotation axes, 163, 240
Allan variance, 55
angular diameter, 228

Baade's star (in Crab Nebula), 123, 126
barycentric correction, 49
beaming factor, 163
binary pulsars, 57, 128
    coalescence, 238, 240
    companions, 135
    disruption, 117
    eclipsing, PSR 1957+20, 133
    evolution, 136
    masses, 134
    orbits, 9, 130, 138
    precession, 60
    PSR 1913+16, 36, 58, 62, 143
    timing, 57
    X-rays, *see under* X-ray binaries
binary stars, 117, 137
birthrate of pulsars, 17, 108, 237
    in supernovae, 110, 240
braking index, 53, 92
bunching, 193, 210
bursters, X-ray, 149

Cassiopeia A, 124, 241
Cen X–3, 2, 141, 145, 148
characteristic age, 53, 105, 109, 128
clocks, standard, 54
coherent radiation, 200, 210
condensed matter, 19, 239
Crab Nebula, 1, 7, 12, 20, 81, 112, 122
    energy supply, 1, 125, 206
    polarisation, 124

    radio source, 124
    scattering, 228
    spectrum, 125, 126
    X-rays, 3
Crab Pulsar, 8, 11, 12, 13, 24, 30, 81, 123, 205
    age, 66
    discovery, 8, 12, 30, 81
    dispersion measure, 41, 90, 91
    double radio pulse, 208
    giant pulses, 30, 82, 88, 210
    glitches, 66
    infrared radiation, 207
    interior oscillations, 74, 78
    optical pulses, 12, 24, 84, 187, 205
    polarisation, 87, 88, 187, 205
    pulse profile, 82, 226
    rotational slowdown, 53, 93
    spectrum, 83, 85
    timing observations, 73, 78, 91
    X-ray pulses, 13, 83, 205
curvature radiation, 188, 192, 197, 205
    critical frequency, 206
    spectrum, 199
cyclotron radiation, 192
cyclotron resonance, 148

death line, 171
density, neutron star, 1, 9
dipole moment, 52
dispersion, 27
    Dispersion Measure, 25, 28, 38, 90, 234
    de-dispersion, 28, 239
distances
    absorption in neutral hydrogen, 39
    dispersion measure, 38, 45
    optical identification, 41, 42
    parallax, 38
Doppler shift, in binaries, 143
drifting pulses, 174
    after a null, 179
    drift rates, 173, 177

eclipsing binary PSR 1957+20, 133, 238
ecliptic coordinates, 47

Index

electron distribution, in Galaxy, 45
ellipticity of neutron star, 74
energy outflow, 52, 80, 127
equation of state, 19
evolution of stars, 114, 137
EXOSAT, 142
extragalactic pulsars, 15, 36, 98

Faraday rotation, 25, 233, 234
frequency derivatives, 53, 92
Fresnel distance, 215, 230

galactic bulge X-ray sources, 149, 152
Galaxy, structure, 45
   electron distribution, 45
   hydrogen distribution, 42
   magnetic field, 235, 242
   pulsar distribution, 100
gamma rays, 24
gamma-ray bursts, 119
gap, magnetospheric, 184, 203
general relativity, 48, 50, 59, 60, 143
glitch, 12, 22, 63, 70, 240
   Crab Pulsar, 66, 92
   Vela Pulsar, 64, 95
glitch function, 75
globular cluster pulsars, 37, 132
globular cluster X-ray sources, 145
gravitational radiation, 9, 61, 140
group velocity, 27
gyrofrequency, 193

Her X–1, 2, 142, 144, 146, 148
hydrogen, distribution in Galaxy, 42
   H II regions, 44
   ionised, 43
   neutral, 39, 42

interplanetary scintillation, 4
interpulses, 160
interstellar medium, 25, 216, 232, 242
interstellar scattering, 35, 211
inverse Compton radiation, 199, 204

Kolmogorov spectrum, 215, 216, 220, 242

Larmor frequency, 192
LMC pulsar PSR 0540−69, 15, 98
low-mass X-ray binaries (LXMBs), 133, 145, 152
luminosity decay
   pulsars, 107
   supernovae, 116
luminosity distribution, 101

magnetic field of pulsars, 20, 53
   decay, 106, 139, 152
   dipole radiation, 52
magnetic field, interstellar, 232, 242
magnetosphere, 21
   charge density, 22
   gaps, 184, 203

maser amplification, 200
mass of neutron stars, 18
mass function, in binaries, 134
mass transfer, 22, 117, 137
massive X-ray binaries (MXBs), 143
mean and true anomaly, 50
microstructure, 24, 168, 181
   bandwidth, 183
   periodic, 183
millisecond pulsars, 11, 18, 36, 51, 128, 238
   formation, 120
   population, 110
   PSR 1937+21, 37
modes, 24, 154, 172
   orthogonal, 181
modulation, radio pulses, 174
moment of inertia, 19, 80

neutral hydrogen, 42
neutron drip point, 20
neutron stars, 7, 11, 18
   thermal radiation, 142
   structure, 19, 150, 239
   surface structure, 150
   surface temperature, 142
neutron superfluid, 20, 76
neutrinos, from supernova, 116
North Galactic Spur, 236
nulls, 24, 170, 174

optical identification of pulsars, 41, 135
optical pulses, 12, 188
oscillations
   of condensed stars, 8
   of Crab Pulsar, 74
outer magnetospheric gaps, 203

pair creation, 204
parallax, 38
periods, 17
   rate of change, 52, 81, 91
planetary orbits, 9
polar cap emission, 165, 184
   geometry, 189
polar magnetic field, 53
polarisation, pulsars, 23, 95, 157
   circular, 164
   optical, 187
   orthogonal, 165
   Stokes' parameters, 155
   subpulses, 181
   synchrotron radiation, 196
polarisation, galactic background, 233
population, 18, 100, 128, 237
   millisecond pulsars, 110
   total in Galaxy, 104
positions, from timing, 46
precursor pulse
   Crab, 83
   Vela, 95, 209
proper motion, 56, 105
   from scintillation, 221

pulse broadening, 89, 110, 223, 225
pulse profiles
  individual, 168
  integrated, 154
  frequency dependence, 159
  polarisation, 157
pulse timing, 46
pulses
  angular width, 156, 162, 168, 190
  intensities, 170

quasi-periodic oscillations (X-ray), 152
quiet collapse, white dwarf, 118, 241

radio pulses, 23
  individual, 168
  integrated, 154
  intensities, 170
  nulls and modes, 24, 154, 170, 174
  subpulses, 173
  theory, 209
radius-to-frequency mapping, 191, 209
Rapid Burster, 151
reference frame, 51
relativistic beaming, 184, 201
Relativistic Binary PSR 1913+16, 58, 62, 143
relativistic corrections, 50
  in binary orbit, 60
Roche lobe, 138
rotation rate limit, 10
Rotation Measure, 234
rotation slowdown, 11, 67, 91

scattering, 211, 224, 229
scintillation, 4, 25, 211
  frequency drifting, 217, 220, 240
  pattern velocity, 221
  refractive, 230
  slow, 89, 229
  thin screen theory, 212
  thick screen theory, 215
Sco X-1, 2, 141, 153
search techniques, 28
searches, 26, 239
self-absorption, 199, 208
SMC Pulsar PSR 0042-73, 16, 36
SMC X-1, 143
spin-up, 146
  upper limit, 139
spectrum, radio, 159, 160, 183
spin-up, in binaries, 18
starquakes, 74
stellar collapse, 114, 240

subpulses, 24, 168, 173
superfluidity, 76
supernova remnants (SNRs), 121, 142, 240
  pulsars in, 98, 110, 121
supernovae, 110, 112
  frequency, 120
  in binary, 116
  SN 1987A, 113, 241
  Types I & II, 113, 114, 115
surveys, 34, 100
  selection effects, 35
synchrotron radiation, 124, 194, 198
  electron lifetime, 125, 196

thermal radiation, neutron stars, 120, 141, 142
  X-ray binaries, 148
  X-ray burster, 150
timing noise, 63, 72
  activity parameter, 72
  starquakes, 74
Tkachenko oscillations, 78

UHURU satellite, 2, 96, 141

Vela Pulsar, 8, 12, 14, 64, 93, 208, 225
  discovery, 93
  gamma-ray pulse, 93, 208
  glitches, 64, 95
  lifetime, 64
  optical pulse, 93, 208
  polarisation, 95
  pulse profile, 93, 225
velocities, 18, 105, 121
velocity-of-light cylinder, 21
vortex lattice oscillation, 78
vorticity in superfluid, 76
  pinning, 77

white dwarf stars, 7, 119, 135, 142
Wolf–Rayet binary, 117

X-ray binaries, 2, 22, 117, 136, 141
  light curves, 141, 148
  masses, 144
  optical companions, 144
  in Galactic clusters, 145
X-ray bursters, 149
X-ray spectral line, 148
X-rays from pulsars, 3, 13, 15, 24, 96

young pulsars, 98

Zeeman effect, 233